Advances in Soil Science

Advances in Soil Science

B.A. Stewart, Editor

Advances in Soil Science

Volume 10

Edited by B.A. Stewart

With Contributions by
E.G. Beauchamp, D. Binkley, R.J. Buresh, S.K. De Datta, S.C. Hart,
M.B. McBride, J.W. Paul, J.T. Trevors, and A. Van Wambeke

With 36 Illustrations

Springer-Verlag
New York Berlin Heidelberg
London Paris Tokyo

B.A. Stewart
USDA Conservation and Production Research Laboratory
Bushland, Texas 79012, USA

Printed on acid-free paper.

ISSN: 0176-9340

Softcover reprint of the hardcover 1st edition 1989

Typeset by Publishers Service, Bozeman, Montana.

9 8 7 6 5 4 3 2 1

ISBN-13: 978-1-4613-8849-4 e-ISBN-13: 978-1-4613-8847-0
DOI: 10.1007/978-1-4613-8847-0

Preface

Soil is formed from physical and chemical weathering of rocks—processes described historically because they involve eons of time—by glaciation and by wind and water transport of soil materials, later deposited in deltas and loessial planes. Soil undergoes further transformations over time and provides a habitat for biological life and a base for the development of civilizations. Soil is dynamic—always changing as a result of the forces of nature and particularly by human influences. The soil has been studied as long as history has been documented. Numerous references to soil are found in historical writings such as Aristotle (384–322 B.C.), Theophrastus (372–286 B.C.), Cato the Elder (234–149 B.C.) and Varro (116–27 B.C.). Some of the earliest historical references have to do with erosional forces of wind and water.

The study of soils today has taken on increased importance because a rapidly expanding population is placing demands on the soil never before experienced. This has led to an increase in land degradation and desertification. Desertification is largely synonymous with land degradation but in an arid land context. Deterioration of soil resources is largely human induced. Poverty, ignorance, and greed are the indirect causes of desertification. The direct cause is mismanagement of the land by practices such as overgrazing, tree removal, improper tillage, poorly designed and managed water distribution systems, and overexploitation. The arrest and possible reversal of land degradation will require policies and technologies for natural resource management. In recent years, there has been a significant shift in emphasis of soil science research away from maximum crop production to the sustainability of crop production systems and to maintaining the quality of soil and water resources. Unless these goals are achieved, land degradation will continue and the land will gradually lose its inherent productivity, threatening the livelihood of those who depend on it.

The focus on increasing food production in recent years throughout much of the world has been on irrigated agriculture. Since 1950, the amount of irrigated land has increased from about 94 million hectares to about 220 million hectares. During the 1980s, however, the rate of irrigation development has dropped materially and is presently less than 1 percent per year. Consequently, it appears certain that much of the additional food production needed in future years must come from the 80 percent of the cultivated land which is not irrigated, and from developing new lands, some of which will surely be fragile lands that can be easily degraded unless careful management practices are implemented.

This series, *Advances in Soil Science*, was established to provide a forum for leading scientists to analyze and summarize the available scientific information on a subject, assessing its importance and identifying additional research needs. This goal seems even more appropriate today than in 1982 when the idea of the series was formulated. There has been much learned about our soil resources. The principles learned and the technology developed need to be used to increase food production and sustain the productivity of the resource base. *Advances in Soil Science* fills a gap between the scientific journal and the comprehensive reference books. Scientists can delve in depth on a particular subject relating to soil science. Contributors are asked in particular to develop and identify principles that have practical applications to both developing and developed agricultures.

Advances in Soil Science was formulated to be international in scope and to cover all subjects relating to soil science. This volume continues in that format in that it contains contributions from scientists from Canada, The Philippines, and the United States on subjects relating to soil classification, nitrogen availability in forest soils, denitrification, nitrogen management in irrigated rice, and heavy metal solubilities in soils. Although we consider our audience to be primarily scientists and students of soil science, the series provides technical information to anyone interested in our natural resources and human influence on these resources. Research in the future will focus on systems that are resource efficient and environmentally sound. The need to optimize crop production while conserving the resource base has never been greater.

The quick acceptance of *Advances in Soil Science* by both authors and readers has been very gratifying and confirms our perception that a need did exist for a medium to publish soil science reviews. I thank the authors for their excellent contributions and cooperation. I also thank members of the Editorial Board for their assistance in selecting such competent authors and the Springer-Verlag staff for their kind assistance and counsel. Finally, and most importantly, I thank the readers for their acceptance and use of *Advances in Soil Science*.

B.A. Stewart

Contents

Contributors

E.G. Beauchamp, Department of Land Resource Science, University of Guelph, Guelph, Ontario N1G 2W1, Canada.

D. Binkley, Department of Forest and Wood Sciences, Colorado State University, Fort Collins, Colorado 80523, U.S.A.

R.J. Buresh, International Rice Research Institute, P.O. Box 933, Manila, The Philippines.

S.K. De Datta, International Rice Research Institute, P.O. Box 933, Manila, The Philippines.

S.C. Hart, Department of Plant and Soil Biology, University of California, Berkeley, California 94720, U.S.A.

M. McBride, Department of Agronomy, Cornell University, Ithaca, New York 14853, U.S.A.

J.W. Paul, Department of Land Resource Science, Ontario Agricultural College, University of Guelph, Guelph, Ontario N1G 2W1, Canada.

J.T. Trevors, Department of Environmental Biology, Ontario Agricultural College, University of Guelph, Guelph, Ontario N1G 2W1, Canada.

A. Van Wambeke, Department of Agronomy, Cornell University, Ithaca, New York 14853, U.S.A.

Reactions Controlling Heavy Metal Solubility in Soils

M.B. McBride*

I. Introduction

Soil chemists have long-recognized that knowledge of the elemental composition of soils is generally of little use in assessing the availability of these elements to plants. An obvious illustration of this principle is the common occurrence of Fe and Mn deficiency in plants despite the relatively high levels of Fe and Mn in many soils. For this reason, chemical soil tests have relied on measurement of extractable or "labile" fractions of elements. Such tests are empirical and provide little basis to relate metal extractability to the chemical forms of the metal in the soil. As soils are increasingly used in our society for purposes other than agriculture, the frequency and extent of soil contamination by toxic metals will increase. Empirical relationships may have to be replaced by a more fundamental understanding of the soil processes controlling metal solubility to prevent practices that could have deleterious effects on soil productivity and environmental quality.

*Department of Agronomy, Cornell University, Ithaca, New York 14853.

Advances in Soil Science, Volume 10

Much of our field and laboratory research on soils is short-term, yet the ultimate fate of metals added to soils can be determined directly only by long-term experiments. It is unrealistic to expect that an adequate number of such experiments will ever be done because an overwhelming range of soil types with different mineralogy and chemistry as well as many different metals would have to be studied. Moreover, chemical interactions among elements that may be very significant in modifying metal solubility have been studied infrequently. It would appear that experimental determination of the fate of metals in soils, although a laudable goal, must be replaced in most cases by the application of general chemical principles to specific situations. This requires a basic understanding of all chemical processes in soils likely to influence metal solubility. This review will attempt to summarize the extent of our knowledge of these processes and their relative importance in affecting metal solubility.

II. Ion Exchange on Layer Silicates

The permanent charge sites of layer silicate clays retain metal cations by nonspecific electrostatic forces and, in the absence of conditions that would favor metal hydrolysis (e.g., high pH), divalent (M^{2+}) and trivalent (M^{3+}) transition and heavy metal cations show typical ion exchange behavior on layer silicates (el-Sayed et al., 1970; McBride, 1976; McBride, 1980c). Both ultraviolet (UV)-visible and electron spin resonance (ESR) spectroscopy have confirmed that ions such as Cu^{2+}, Co^{2+}, Ni^{2+}, Mn^{2+}, VO^{2+}, and Cr^{3+} retain their inner hydration sphere as well as a high degree of rotational mobility on smectite exchange sites, offering direct support for the involvement of electrostatic forces only (Clementz et al., 1973; McBride et al., 1975; McBride, 1979b; Schoonheydt, 1982). Thus, strength of metal bonding should depend only on the charge and hydration properties of the cation. However, if the interlamellar spacing of the clay is limited to the equivalent of one or two molecular layers of water, a high degree of motional restriction of the hydrated metal ions is observed. For example, in vermiculites, M^{2+} ions are oriented in the interlayer as hexa-aquo complexes under both wet and air-dry conditions because the interlayer does not expand sufficiently in water to remove the steric restraint to ion motion.

Ion exchange selectivity diagrams (e.g., see Figure 1) suggest a very high preference of clays for higher-charge cations at low ionic strength. However, this high selectivity does not arise from a specific bonding mechanism, but is a consequence of the mathematical form of the ion exchange equation. Consider, for example, Cu^{2+}/Na^{+} exchange:

$$Cu^{2+} + 2\ Na^{+}\text{-clay} \rightleftharpoons Cu^{2+}\text{-clay} + 2\ Na^{+}$$

The selectivity coefficient can be written:

$$K_s = \left(\frac{(m_{Na})^2 \cdot N_{Cu}}{m_{Cu} \cdot (N_{Na})^2}\right) \quad (1)$$

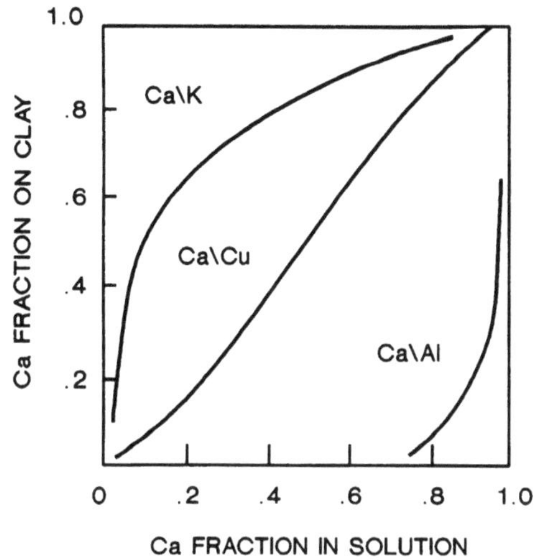

Figure 1. Selectivity curves for Ca^{2+} exchange at low ionic strength with K^+, Cu^{2+}, and Al^{3+} on smectite. Equivalent fractions of Ca^{2+} on exchange sites and in solution are plotted. (Data from McBride and Bloom, 1977; el-Sayed et al., 1970; and Hutcheon, 1966).

where m_{Na} and m_{Cu} represent the molarity of the metals in the solution phase and N_{Na} and N_{Cu} represent the fraction of clay exchange sites occupied by these ions. If the exchange cations behave as an ideal "solution" in the clay phase, then K_s should remain relatively constant with changing ionic strength (Sposito, 1981). As a result, lowering the ionic strength of the clay suspension by water dilution would necessarily lower $(m_{Na})^2/m_{Cu}$), and the exchange complex would compensate for this by adsorbing more Cu^{2+} at the expense of Na^+ (i.e., increasing $N_{Cu}/N_{Na}{}^2$). This is the well-known concentration-charge effect of ion exchange theory, and it can produce high *apparent* preferences for high-charge ions on exchange sites. As the calculated isotherms—(assuming equal preference of Ca^{2+} and Na^+ for sites)—reveal in Figure 2, low ionic strength necessarily results in the efficient partitioning of the higher-charge ion onto the exchange sites. In general, it can be demonstrated that if a mass-action ion exchange equation of the type described by equation 1 is valid, every different electrolyte concentration *and* clay concentration will produce a different adsorption isotherm of the type plotted in Figure 2. However, this figure illustrates the effect of electrolyte concentration at constant clay concentration only.

An additional complicating factor in quantifying the affinities of metals for clays is the "particle concentration" effect. This effect can be illustrated by equilibrating a solution of metal ions with clay, then removing a portion of the solution phase from the clay suspension by filtration. In principle, the composition of the clay and solution phases should be unaffected by concentrating the clay in this manner. However, desorption of Co^{2+} and Ni^{2+} from montmorillonite has been observed upon concentrating the clay, an indication that the partition coefficient (i.e., adsorption affinity) of the metal is reduced by increasing the solids concentration (DiToro et al., 1986). Although such effects might be discounted as caused by experimental artifacts or overlooked chemical components

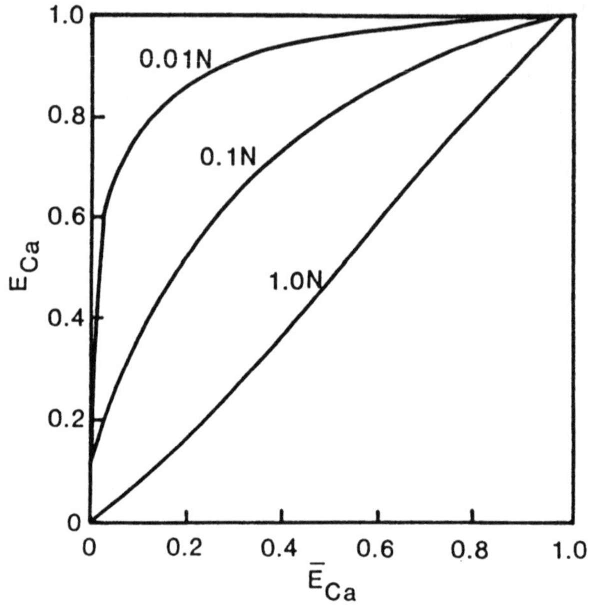

Figure 2. Calculated Ca^{2+}/Na^{+} exchange isotherms at three different ionic strengths. Equivalent fractions of Ca^{2+} in solution (E_{ca}) and on the exchanger (E_{ca}) are plotted. (Sposito, 1981; reprinted with permission of Clarendon Press, Oxford, UK.)

in solution, they have appeared in numerous adsorbate/adsorbent systems and may require a more general explanation. One suggestion is that interparticle associations somehow induce the desorption of metals (DiToro et al., 1986). As smectite particle associations depend on particle concentration, any influence of particle association on cation selectivity will produce a "particle concentration" effect. Apparent nonreversibility in ion exchange processes between monovalent and divalent cations on smectites has been attributed to slow and imperfect dispersion of flocculated suspensions (van Bladel and Laudelout, 1967), with the effect that the selectivity for the divalent metal appears greater during desorption than during adsorption of this metal. It remains to be established whether the "particle concentration" effect will be fully explained by particle association phenomena. One cannot rule out the possibility that existing models of ion exchange are inadequate and that metal ion activities in the solid phase are not fully described by mole or equivalent fractions on exchange sites, even after adjustments for nonideal behavior. It may be necessary to include clay/solution ratios as parameters in the description of ion exchange.

In the soil environment, heavy metals such as Cu^{2+}, Cd^{2+}, and Pb^{2+} must compete with Ca^{2+}, Mg^{2+}, and other more abundant cations for exchange sites. Therefore, strong partitioning of the heavy metals onto exchange sites is not expected. Consider, for example, Cu^{2+}/Ca^{2+} exchange under acidic conditions, the selectivity coefficient is nearly unity (el-Sayed et al., 1970), that is,

$$K_s = \left(\frac{m_{Ca}}{m_{Cu}}\right)\left(\frac{N_{Cu}}{N_{Ca}}\right) = 1 \qquad (2)$$

where the symbols are defined in an analogous manner to those used in equation 1. Typical Cu contents and cation exchange capacities of clay fractions from soils of the northeastern United States are about 100 mg/kg and 250 mmole/kg, respectively (L.J. Johnson and Chu, 1983). Estimating the level of soluble Ca^{2+} at about 10^{-3} *M* and assuming that all soil Cu^{2+} is exchangeable, one would estimate a Cu^{2+} solubility in these soils to be about 10^{-5} *M* from equation 2. The much lower solubilities obtained in reality (Sanders, 1982) lead to the obvious conclusion that mechanisms much more specific than ion exchange are needed to explain the behavior of Cu and other heavy metals. In fact, evidence for specific adsorption of metals such as Co^{2+}, Zn^{2+}, and Cd^{2+} (Hodgson, 1960; Tiller and Hodgson, 1962; Garcia-Miragaya and Page, 1976, 1977) has been obtained for very low adsorption levels on layer silicate clays. This may indicate the presence of a few sites capable of chemisorbing these metals, possibly -SiOH or -AlOH groups at clay edges (Inskeep and Baham, 1983) or sites associated with oxide and organic impurities in the clays (Jenne, 1968). Some researchers have concluded that only reversible ion exchange of metals such as Co^{2+} and Zn^{2+} is operative on montmorillonite at pH values below 6 (Maes and Cremers, 1975) and that the greater specificity of layer silicate clays for transition and heavy metals arising at higher pH (Farrah and Pickering, 1976, 1977; Tiller et al., 1979) is explained by the formation of hydrolysis products. The irreversibility of adsorption at higher pH is attributed to strong adsorption of hydroxy polymers at the silicate surface (Hodgson et al., 1964; Farrah and Pickering, 1976; Tiller et al., 1984). Transition and heavy metals possess a much greater tendency than the alkaline earth metals to hydrolyze because of their electronic structure. Clays added to solutions of transition and heavy metals generally induce a lowering of the pH, which is attributed to enhanced hydrolysis of the metals (Ragland and Coleman, 1960; Bloom et al., 1977; McBride, 1982c), but direct identification of adsorbed hydrolysis products has been difficult. Based on ESR experiments with Cu^{2+} and Cr^{3+} (the fully hydrated cation produced a characteristic spectrum that is altered by hydrolysis) it appears that in the low pH range smectites promote hydrolysis beyond that observed in aqueous solution. Conversely, at high pH, complete hydrolysis of the metal to form the neutral metal hydroxide appears to be inhibited. These phenomena can be understood when it is realized that layer silicate clays have a high affinity for polymeric hydroxymetal cations (Rengasamy and Oades, 1978; Hodges and Zelazny, 1983). Thus, the first step of the hydrolysis sequence

$$xM(H_2O)_6^{n+} \rightleftharpoons M_x(OH)_y^{(nx-y)^+} \rightleftharpoons M(OH)_n^0$$

is promoted by layer silicates, probably because the polycations interact intimately and extensively with the siloxane surfaces via the hydroxyl groups. Hydrated metal cations, on the other hand, are only loosely bonded to the surfaces by electrostatic forces (as indicated earlier). Strong bonding of polycations would necessarily obstruct the second step of this reaction, preventing complete hydrolysis to form the neutral hydroxide. For example, numerous studies have shown that gibbsite formation is prevented or delayed in clay systems where Al^{3+} is adsorbed on smectite despite the fact that the $(Al^{3+})(OH^-)^3$ ion activity product often exceeds the solubility product of gibbsite (Turner and Brydon, 1965, 1967). The affinity of the charged hydroxymetal polymers for the silicate surface probably presents an energy barrier to nucleation and precipitation of a separate hydroxide phase.

Based on available evidence, it is unlikely that clay exchange sites limit trace metal solubility in soils. More selective mechanisms are indicated by the low solubility and difficult extraction of these metals. Nevertheless, in metal-polluted soils, exchange sites have an important role in attenuating the movement of these metals through the soil.

III. Chemisorption on Mineral Surfaces

There is no lack of evidence that transition and heavy metals in soils when present at trace levels are retained largely in nonexchangeable forms. Schemes for complete metal extraction require extreme treatments, including the oxidative degradation of organic matter and dissolution of Fe and Mn oxides (Shuman, 1979). Even the preferential adsorption of polymeric hydroxymetal cations by layer silicates would not seem to account for the stability of these sorbed forms of metal.

It is now established that metal oxides and hydroxides as well as amorphous aluminosilicates provide surface sites for chemisorption of heavy metals. Indirect evidence for the formation of surface-metal bonds includes:

1. The release of as many as two H^+ ions for each M^{2+} ion adsorbed (Forbes et al., 1976).
2. The high degree of specificity shown by oxides for particular metals (Kinniburgh et al., 1976).
3. Changes in the surface charge properties of the oxide as a result of adsorption (Stumm and Morgan, 1981).

The last effect is attributed to the increased surface positive charge developed by chemisorption. For example, on goethite, the proposed surface reaction is

$$\text{-Fe-OH} + M(H_2O)_6^{2+} \rightarrow \text{-Fe-O } M(H_2O)_5^{+} + H_3O^{+}$$

and can be detected as a change in electrophoretic mobility of suspended oxide particles. For example, in Figure 3, Mg^{2+} adsorption on goethite prevents the oxide from developing a negative charge at high pH (Bleam and McBride, 1985), thereby shifting the effective zero point of charge (ZPC) to a much higher pH.

More direct evidence for surface coordination of metals has been obtained from ESR and UV-visible spectra of the adsorbed ions (Martini et al., 1980; Clark and McBride, 1984; McBride et al., 1984; McBride, 1985a; Bleam and McBride, 1986). The UV-visible spectra of transition metals are sensitive to

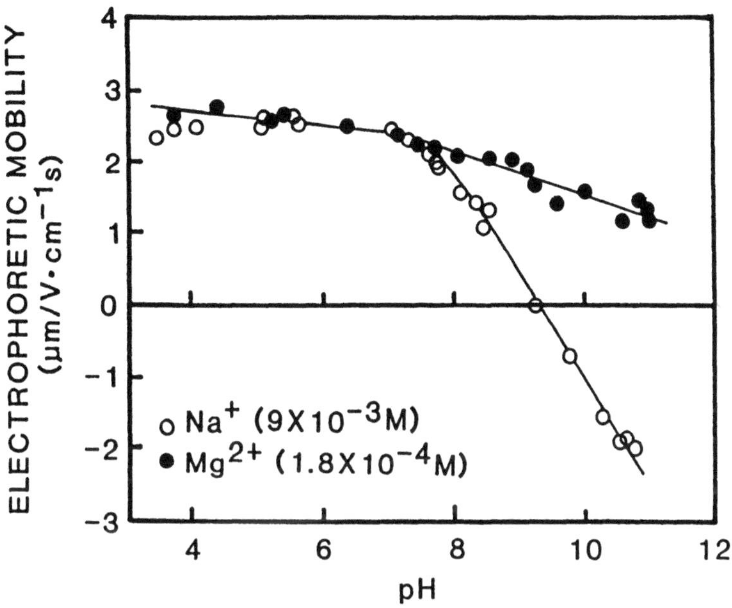

Figure 3. Variation of electrophoretic mobility of microcrystalline goethite suspensions as a function of pH in 9×10^{-3} *M* $NaClO_4$, with and without Mg^{2+} present. (Adapted from Bleam and McBride, 1985.)

inner-sphere coordination by ligands because replacement of H_2O in the aquometal complex by another ligand alters the crystal field-splitting energy of the d-orbitals. This is demonstrated for Cu^{2+} bonded on Al hydroxide in Figure 4, where a shift in the visible absorption spectrum observed upon bonding is interpreted as an increase in the separation of d-orbital energy levels. Such a result would suggest that surface oxyanions are stronger ligands (in the sense of being more effective Lewis bases) than water molecules. However, ambiguity exists in this interpretation because the formation of hydroxymetal species at the surface would also produce a change in UV-visible spectra. The ESR spectrum of Cu^{2+}, which is also sensitive to the coordination environment of the metal ion, is altered on adsorption, with the *g*-value decreasing and the hyperfine splitting (*A*-value) increasing relative to the free $Cu(H_2O)_6^{2+}$ ion (McBride, 1982a). This shift can be interpreted as increased covalency of ligand-metal bonding (Kivelson and Neiman, 1961) or as an increase in crystal-field splitting. Assuming that Cu^{2+} remains in its preferred tetragonally distorted octahedral coordination on adsorption and that the spin-orbit coupling constant, ζ, is unchanged by adsorption, then the principal components of the *g*-tensor are given by the equations (Carrington and McLachlan, 1967):

$$g_{11} = 2\left(1 + \frac{4\zeta}{\Delta_1}\right) \tag{3}$$

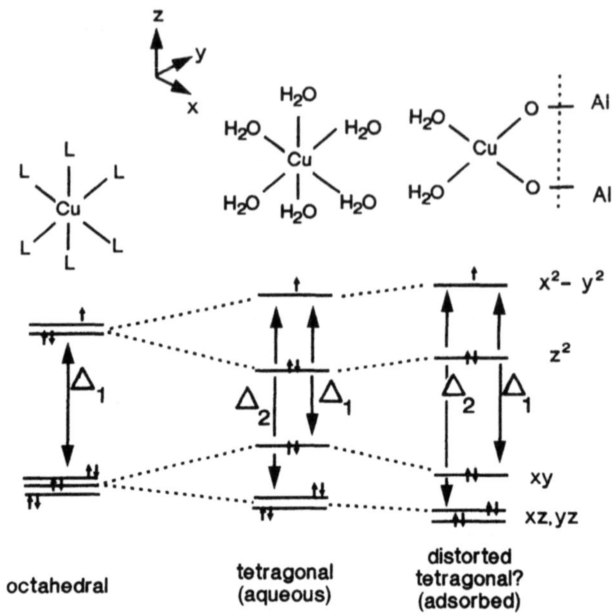

Figure 4. Diagram of d-orbital energy levels for octahedral $Cu(H_2O)_6{}^{2+}$, tetragonal $Cu(H_2O)_6{}^{2+}$, and oxide-bound Cu^{2+}.

$$g_{\perp} = 2\left(1 + \frac{\zeta}{\Delta_2}\right) \tag{4}$$

Here, the Δ_1 and Δ_2 crystal-field splitting energies are defined in Figure 4; g_{11} and $g\hat{\perp}$ are the g-values for the unpaired electron in Cu^{2+} for the case of the external magnetic field, H, applied parallel and perpendicular, respectively, to the symmetry or z axis of Cu^{2+}. An increase in Δ_1, the energy separation of the d_{xy} and $d_{x^2+y^2}$ orbital, would decrease g_{11} as observed. But again, ambiguity arises from the fact that hydroxy–Cu complexes have smaller g_{11} and larger hyperfine splitting than the hydrated metal (McBride, 1982a). The possibility must again be entertained that the change in chemical environment of metals on adsorption on oxides is a consequence of surface-induced hydrolysis in addition to, or rather than, chemisorption. The hydrolysis step is integrated into some oxide-adsorption models because it accounts for the strong correlation between the tendency of metals to hydrolyze and their tendency to adsorb on oxides (R.O. James and Healy, 1972b).

It is clearly proven by ESR that oxide-sorbed metals are located at isolated surface sites, except at high metal loadings, and are bound rigidly in comparison to metal ions sorbed on exchange sites of layer silicate clays. Although these facts alone do not prove the existence of direct metal-oxide bonds, further observations are indicative of chemisorption. For example, noncrystalline Al hydroxides and allophanes adsorb large amounts of Cu^{2+} at discrete surface sites in comparison to microcrystalline gibbsite or boehmite (Clark and McBride, 1984; McBride

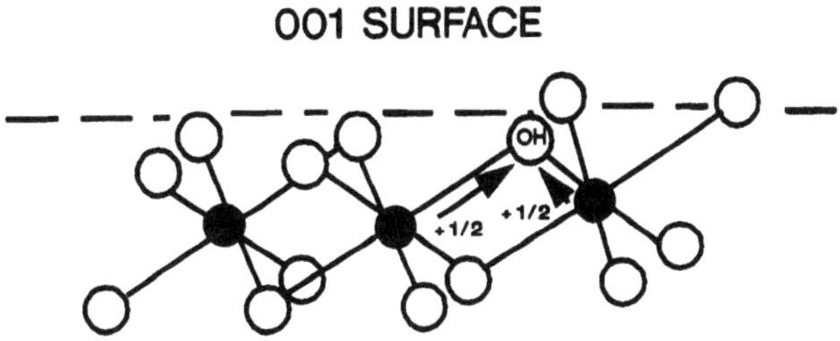

Figure 5. Cross-section of the (001) surface of gibbsite depicting charge-balanced OH groups.

et al., 1984; McBride, 1982a). The dominant (001) surfaces of gibbsite and boehmite are believed to be inactive in chemisorption because the OH groups of these surfaces are coordinated to at least two Al^{3+} ions. Figure 5 reveals that these "bridging" OH groups at the (001) plane of gibbsite are electrostatically satisfied and have no tendency to further coordinate with metal cations. In contrast, "edges" and defects on surfaces of noncrystalline oxides (and to a lesser extent on crystalline oxides) possess OH groups coordinated to single Al atoms. The oxygen atoms of these groups do not have their formal charge fully balanced by structural cations, so that they tend to protonate:

$$-Al-OH]^{-1/2} + H^+ \rightleftharpoons -Al-OH_2]^{+1/2}$$

Alternatively, the oxygen charge can be compensated by a multivalent cation such as Cu^{2+}:

$$-Al-OH]^{-1/2} + Cu^{2+} \rightarrow -Al-O-Cu]^{+1/2} + H^+$$

Some additional discussion of the behavior of these "bonding" groups is warranted. The singly coordinated OH^- groups are prevalent on certain crystal surfaces of oxides, for example, the (100) face of goethite (Russell et al., 1974), and are believed to be responsible for chemisorption of anions (such as phosphate) and metals. Figure 6 provides a diagrammatic rationale for the chemical behavior of these bonding groups based on the electron-withdrawing power of the cations coordinated to the surface oxyanion. The degree of electron polarization away from the bonding lone-pair groups of the O atom, estimated from the electronegativity of the coordinated H^+ and Al^{3+}, provides a basis for predicting the tendency of differently coordinated oxyanions to bond covalently with metals. For example, an oxyanion coordinated to one Al^{3+} and one H^+ (Figure 6b) is likely to form a bond with metals, whereas one coordinated to two Al^{3+} and one H^+ (Figure 6a) is not. Thus Cu^{2+} bonds to the former type of Al-OH group, generally found on crystal edges, but not to the latter OH groups, which reside on dominant crystal surfaces such as the (001) plane of gibbsite. By similar reasoning, although the bridging OH (Figure 6a) is likely to be more acidic than the bonding OH (Figure 6b), releasing its proton more readily, the bonding OH would be much more likely to accept an additional proton in acidic solution, thus forming

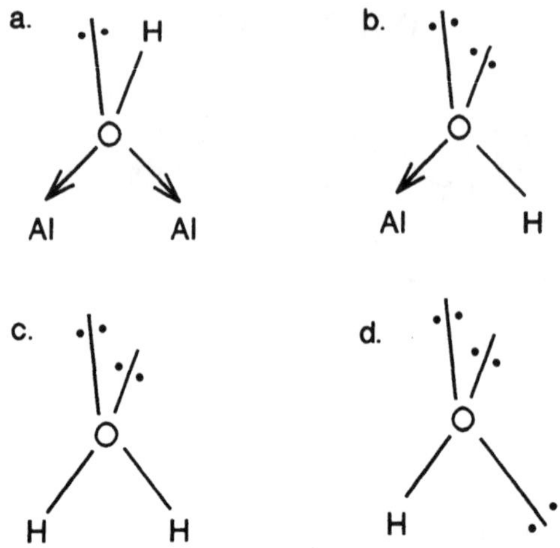

Figure 6. Schematic diagram of nucleophilic oxyanions in different chemical environments: (a) bridging OH of oxide, (b) bonding OH of oxide, (c) H_2O, (d) OH^-.

a positively charged Al-OH_2^+ site. This site is conducive to ligand exchange by phosphate and other oxyanions since H_2O is a good "leaving group." Thus, the interesting conclusion is reached that the same oxide surface sites are responsible for metal cation and anion adsorption and that these sites have the highest surface density on poorly crystalline oxides and aluminosilicates.

Extending this reasoning to water and OH^- as ligands for metal coördination complexes (Figures 6c, 6d), one would predict that transition metal-hydroxy complexes are intrinsically more stable than metal-water complexes. This prediction appears to be confirmed by the fact that the ESR spectrum of $Cu(OH)_4^{2-}$ has larger hyperfine splittings and smaller *g*-values than the spectrum of $Cu(H_2O)_6^{2+}$ (Clark and McBride, 1984), indicating that OH^- produces a more covalent bond (stronger crystal field) than H_2O.

The conclusion reached from this discussion is that chemisorption should depend on the degree of crystallinity and surface morphology of the adsorbent. This has been experimentally confirmed, with adsorption at relatively low metal concentrations involving the formation of metal-oxyanion bonds rather than the precipitation of solid phases at the surface.

A further indication of direct surface bonding is the evidence that metal ions may be simultaneously coordinated to the oxide and to a strongly complexing ligand. These ternary complexes include oxide-Cu^{2+} ammonia and oxide-Cu^{2+} phosphate associations (Davis and Leckie, 1978; Bourg et al., 1979; Clark and McBride, 1984, 1985; McBride, 1985a, 1985b). If the metal ion were not strongly bonded, one would expect some of these complexing ligands to "strip" the metal from the surface.

It is very unlikely that all oxides bond the heavy metals by a similar mechanism. Ultraviolet-visible spectroscopy and other less direct techniques indicate that Ni^{2+} and Co^{2+} are adsorbed on silica as the hydrated (hexa-aquo) metal, in contrast to their behavior on Al and Fe oxides (Healy et al., 1968; Hathaway and Lewis, 1969a, 1969b). Confirming evidence from ESR spectra of Cu^{2+} adsorbed on silica at low pH indicates that the adsorbed cation is largely coordinated to H_2O (von Zelewsky and Bemtgen, 1982; Clark and McBride, 1984). Nevertheless, ESR further reveals that the bound ion is rigid, which is indicative of a direct bond between Cu^{2+} and one or more silanol groups, that is,

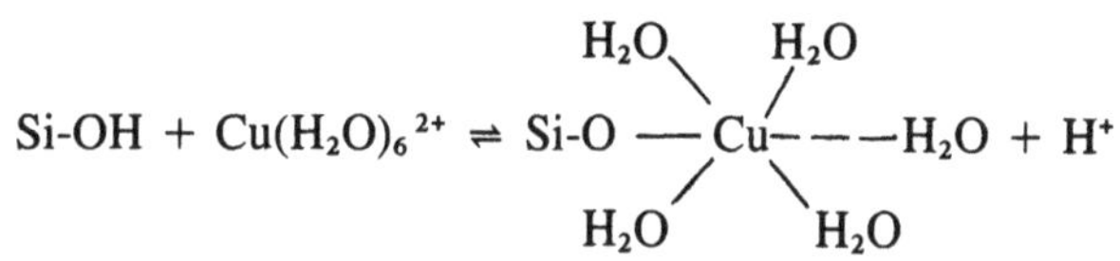

The relatively weak nature of this bond is suggested by the fact that the ESR spectrum of the chemisorbed Cu ions diminishes in intensity as the pH is raised to 6 or higher, an indication that the surface complex is unstable relative to $Cu(OH)_2$:

$$\equiv\text{Si-O-Cu}^+ + \text{NaOH} \rightleftharpoons \equiv\text{Si-O}^-\,\text{Na}^+ + \text{Cu(OH)}_2$$

Similar results have been observed for Cu^{2+} adsorption by imogolite, a paracrystalline aluminosilicate.

It is concluded, then, that aluminol (Al-OH) groups coordinate Cu^{2+} and other transition metals more strongly than do silanol (Si-OH) groups. One can apply a simple electrostatic argument that the surface oxygen atoms of silica, unlike those of alumina, are not strongly inclined to coordinate with multivalent cations because their charge is satisfied:

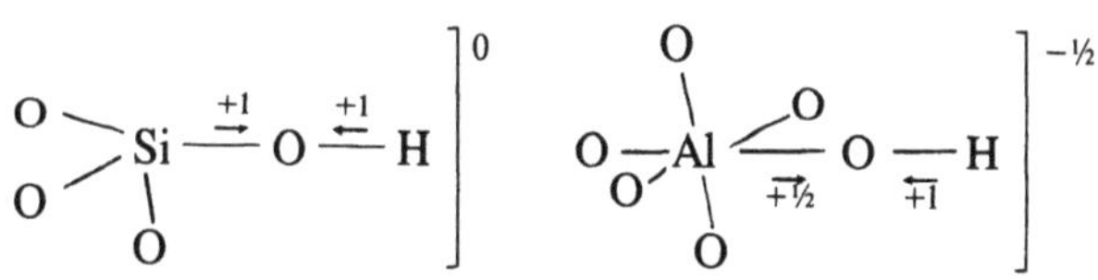

Silanol groups are more acidic than aluminol groups, a fact that can be rationalized on the basis of the strong electrical field of Si^{4+} compared to Al^{3+}, which distributes electron density toward Si and weakens the O-H bond. An equivalent result is predicted by using the argument that Si has a higher electronegativity than Al. Therefore, silanol groups dissociate more readily than aluminol groups, but they coordinate less readily with transition metals.

An additional factor that favors metal bonding to alumina may be the potential for bidentate coordination at crystal edges, depicted diagrammatically in Figure 7. Two chemically distinct coordination sites for Cu^{2+} have been evident in ESR spectra of allophanes, imogolites, and Al hydroxides (McBride, 1982a; Clark and McBride, 1984). The ESR parameters of Cu^{2+} at the two sites detected in allo-

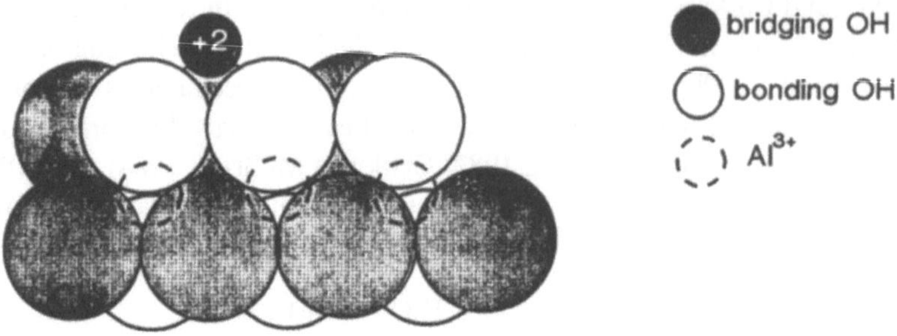

Figure 7. Diagram of a M^{2+} cation bonded at adjacent OH groups on Al hydroxide.

phane are presented in Table 1 along with the estimated crystal-field splittings obtained from equations 3 and 4, assuming that $\zeta = 829\ cm^{-1}$. Site 2 can be visualized as coordinating more strongly with Cu^{2+} than site 1, increasing the splitting of the d-orbital energy levels. As adsorption on allophanes is increased, Cu adsorption at site 1 increases with respect to adsorption at site 2, an expected result if site 2 retained the metal by an energetically favored bidentate bond. The release of 2 H^+ for each divalent heavy metal ion adsorbed on Fe oxides (Forbes et al., 1976; Perona and Leckie, 1985) may indicate the formation of bidentate bonds on oxides as well as allophane.

Various oxides have the ability to chemisorb metals. The Mn oxides are notable for their unusually high selectivity for certain metals such as Pb^{2+}, Co^{2+}, Cu^{2+}, and Ni^{2+} (D.J. Murray et al., 1968; Gadde and Laitinen, 1974; J.W. Murray, 1975; McKenzie, 1980; Golden et al., 1986). It is presumed, based on the pH dependence of metal adsorption, that metal ions are retained by direct coordination to surface oxygens:

$$>Mn^{IV}-OH + Co^{2+} \rightarrow >Mn^{IV}-O-Co^{+} + H^{+}$$

but the process is evidently more complex than this for some metals. Release of Mn^{2+} from the solid accompanies adsorption of some heavy metals (McKenzie, 1970; Traina and Doner, 1985a, 1985b). In certain cases, this may be a consequence of ion exchange of preexisting Mn^{2+} from the oxide surface. Many of the synthetic Mn oxides are prepared at very high pH and, when introduced into acid aqueous media, are thermodynamically unstable. For example, γ-MnOOH, once placed in acidic solution, disproportionates (Bricker, 1965):

$$2\gamma\text{-}MnOOH + 2H^{+} \rightarrow MnO_2 + Mn^{2+} + 2H_2O$$

Some of this Mn^{2+} could be retained on surface charge sites and released on the adsorption of other metals. Adsorption of Co^{2+} seems especially effective in releasing Mn^{2+}, a fact that has been attributed to an electron transfer between the adsorbed Co^{2+} and the oxide (McKenzie, 1970; Traina and Doner, 1985b):

Table 1. Measured ESR parameters and calculated crystal-field splittings for Cu^{2+} bound to allophane

Site	g_{11}	$g_{\perp}$	$A_{11}(cm^{-1})$	$A_{\perp}$	$\Delta_1(cm^{-1})$	$\Delta_2(cm^{-1})$
1	2.362	2.082	0.0129	[a]	18,300	20,200
2	2.336	2.065	0.0156	[a]	19,700	25,500

[a] The hyperfine splitting for the magnetic field aligned perpendicular to the z-axis of Cu^{2+} was not resolved, but was estimated by spectral simulation to be about 0.0015 cm^{-1}.

$$Co^{2+} + MnOOH = CoOOH + Mn^{2+}$$

Darkening of the Mn oxide accompanies adsorption of Co^{2+} and Mn^{2+}, suggesting a change in oxidation state of the mineral (McKenzie, 1980). X-Ray photoelectron spectroscopy (XPS) has confirmed the oxidation of Co^{2+} to Co^{3+} at the surface of birnessite (Dillard and Schenck, 1986). It has been suggested that Co^{3+} has a unique ability to substitute for Mn in the oxide structure (McKenzie, 1970). In addition, XPS analyses of several Mn oxides show Mn(III) to be the dominant oxidation state at the surface (J.W. Murray et al., 1985), suggesting that structural Mn^{3+} can be readily displaced by Co^{3+}. The fact that Co^{2+}, unlike many other metals, has a relatively stable higher oxidation state may in itself enhance adsorption because the Co^{3+} ion should have stronger adsorption tendencies than the Co^{2+} ion. In other words, oxidation of Co^{2+} by Mn oxides could promote the subsequent adsorption of Co, because of the greater affinity of Co^{3+} for oxide surfaces, without actual substitution into the structure. The tendency for soil Co to be concentrated in Mn oxide particles (Taylor, 1968) is possibly a consequence of these surface interactions.

The high apparent adsorption capacity of Mn oxides for Mn^{2+}, Fe^{2+}, and (as mentioned earlier) Co^{2+} may result from precipitation of these metals on oxidation. Mn oxides are known to promote the oxidation of both Mn^{2+} (autocatalysis) and Fe^{2+} (Hem, 1978; Stumm and Morgan, 1981). Adsorption and surface precipitation are not easily distinguishable (Sposito, 1986), so that the close association of Mn with Fe (and several other heavy metals) may be one of coprecipitation rather than adsorption. The possibility that solid solutions, or coprecipitates, control metal solubility will be discussed later in this chapter. It appears that some metals, such as Pb^{2+}, adsorb strongly on Mn oxides without any accompanying oxidation process (McKenzie, 1980). In such cases, accumulation of metals in soil Mn oxides would appear to result from the strength of the covalent bond formed at surface sites.

Further direct evidence for the nature of chemisorbed metal ions on oxides has been obtained in an ESR study of Cu^{2+} on titanium dioxide (Bleam and McBride, 1986). The Cu^{2+} sorption curves reveal two distinct adsorption events, one near pH 2.0 to 3.0 and another near pH 4.5 (Figure 8); these events are characterized by ESR spectra of bound Cu^{2+} that shift from low g-values at low pH to higher

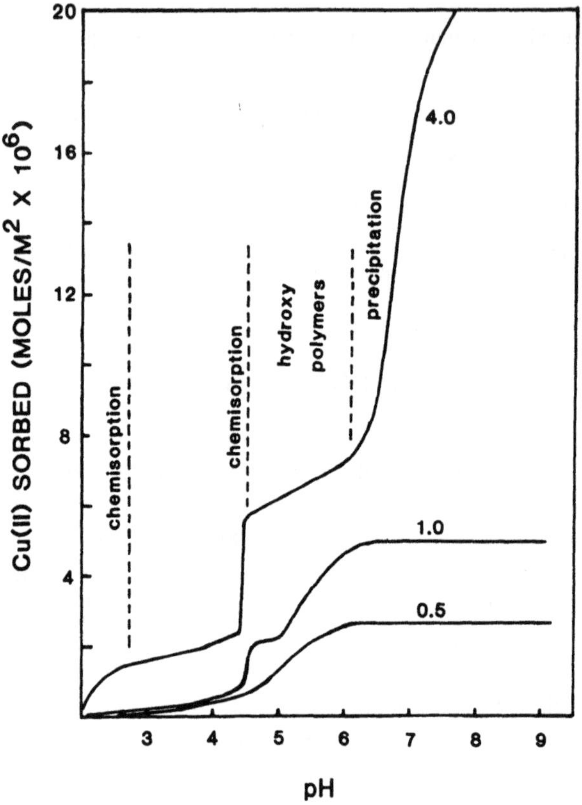

Figure 8. Cu^{2+} sorption on TiO_2 suspension as a function of pH and at three levels of Cu^{2+} addition (equivalent to 0.5, 1.0, and 4.0 monolayers at the surface). (Adapted from Bleam and McBride, 1986.)

g-values at pH 4.5. The interpretation, using arguments similar to those developed for the Cu-Al oxide studies, is that Cu^{2+} chemisorbs strongly at low pH to sites that retain it by a bidentate mechanism, whereas at higher pH a weaker complex is formed with single Ti-OH groups.

Even though titania has a higher ZPC than silica, it adsorbs heavy metals at least as well as silica, a fact that has been attributed to "specific interactions" between the metal ions and TiO_2 (R.D. James and Healy, 1972a). Since Ti is less electronegative than Si, surface Ti-O groups may be stronger Lewis bases than Si-O groups, forming more covalent bonds with metals.

As the pH is raised, adsorption on oxides inevitably merges into precipitation, although this is usually obscured by the apparent continuity of the sorption curve. The uppermost curve in Figure 8, however, clearly reveals an adsorption plateau separated from precipitation above pH 6. The fact that chemisorption and nucleation or precipitation phenomena are resolved in this case—when they generally have not been separated in the case of sorption on oxides of Fe and Al—may be a consequence of the low adsorbate/adsorbent ratios utilized, so that precipitation was suppressed until relatively high pH. Considering that chemisorption can be visualized as a two-step process in which the surface site must dissociate in order for the metal to bond, the ease of proton dissociation from Ti-OH groups compared to the weakly acidic Fe-OH or Al-OH groups may assist metal bonding. Thus, chemisorption on titania is possible at quite low pH.

Oxide surfaces possess selectivity for certain heavy metals and although the order of preference is roughly consistent with the tendency of the metals to hydrolyze (Gerth and Brümmer, 1983), there are notable exceptions. Figure 9 depicts adsorption of metals on Mn and Fe oxides; even hematite and goethite show some difference in selectivity despite the common view that the hydrated surfaces of these two minerals should not behave very differently. Metal affinities for amorphous Fe hydroxides have been reported to follow the order (Kinniburgh et al., 1976):

$$Pb^{2+} > Cu^{2+} > Zn^{2+} > Ni^{2+} > Cd^{2+} \geq Co^{2+} > Sr^{2+} > Mg^{2+}$$

with Al hydroxide producing a somewhat different sequence:

$$Cu^{2+} > Pb^{2+} > Zn^{2+} > Ni^{2+} > Co^{2+} \geq Cd^{2+} > Mg^{2+} > Sr^{2+}$$

Metal affinity for the silanol groups of silica follows the order (Dugger et al., 1964; Schindler et al., 1976):

$$Pb^{2+} > Cu^{2+} > Co^{2+} > Zn^{2+} > Ni^{2+} = Cd^{2+} > Sr^{2+} > Mg^{2+}$$

Clearly, the oxide-metal bond is not entirely electrostatic; otherwise, the measure of "ionic potential" (Z^2/r), based on the charge (Z) and radius (r) of the ion, would have some predictive value. Table 2 demonstrates that electrostatic factors alone would place Mg^{2+} high and Pb^{2+} low on the affinity sequence. The covalent contribution (estimated from electronegativity or "softness" parameters) to bonding is not the overriding factor either, because this would result, for example, in a higher affinity for Cd^{2+} than for Zn^{2+}. It is concluded that the transition metals classified as "hard" (according to the Pearson terminology for acids and bases) are bonded more strongly than the "soft" transition metals in accordance with the Irving–Williams series (see Table 2). However, "soft" *nontransition* metals (e.g., Pb^{2+}) are preferred over harder nontransition metals (e.g., Cd^{2+}, Mg^{2+}). Because Zn^{2+} is a borderline metal—harder than the heavier posttransition metals but softer than the alkaline earth metals—it tends to display intermediate behavior. A high ionic potential causes it to adsorb more strongly than Cd^{2+}, but Zn^{2+} is insufficiently soft to adsorb as strongly as Pb^{2+} or Cu^{2+}. Any predictive model of bonding to oxides will evidently require at least two parameters: an estimate of the electrostatic and the electron-sharing properties of metals.

Models of adsorption on amphoteric surfaces generally fall into one of two groups: the "surface complexation" models, in which adsorption is described in terms of metal exchange of protons from oxyanions (Schindler et al., 1976;

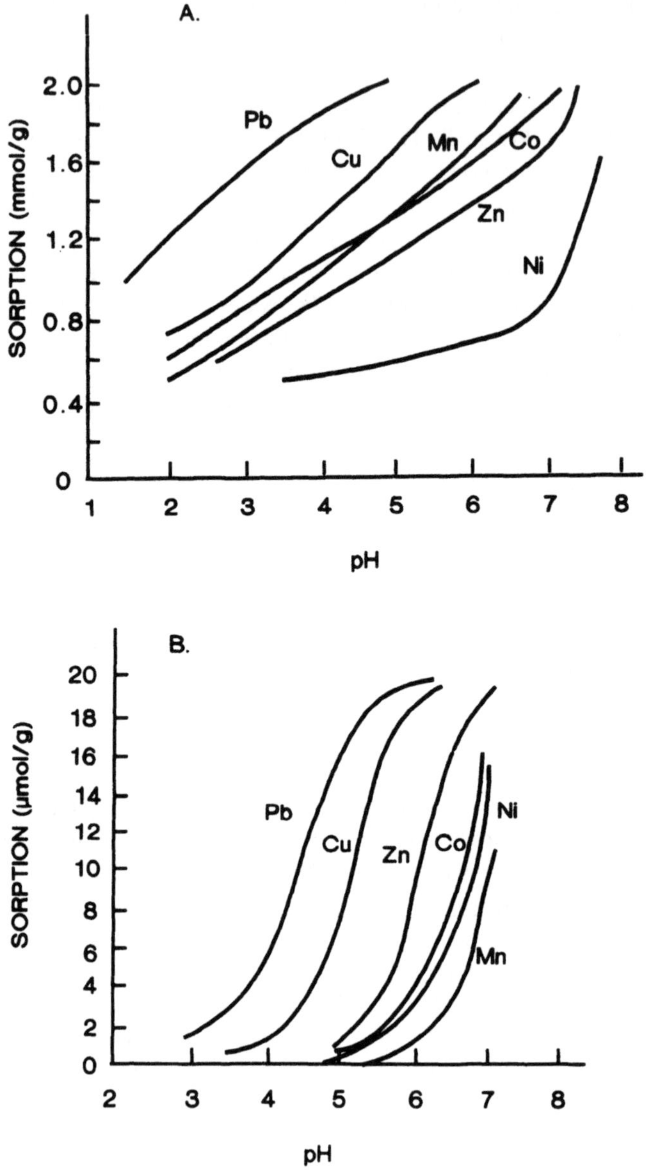

Figure 9. Metal adsorption versus pH on (a) Mn oxide (birnessite), (b) hematite, (c) goethite. (Reproduced from McKenzie, 1980, with permission.)

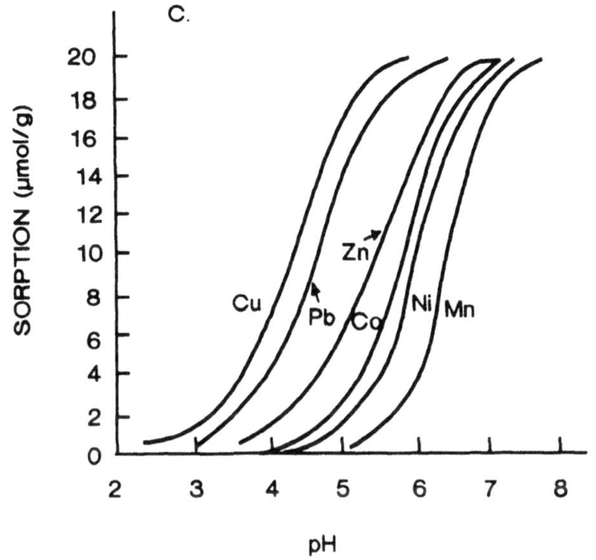

Figure 9. *Continued.*

Stumm et al., 1976) or the "diffuse multiple layer" models, in which adsorption occurs in response to the pH-dependent electrical potential at the surface (Bowden et al., 1977). Hybrids of these models also exist; several of the more prominent ones have been compared by Westall and Hohl (1980). Unmodified diffuse double-layer theory would allow no chemisorption, with metal adsorption possible only at pH values in the vicinity of, or higher than, the ZPC, which is an approximate function of the ionic potential of the metal in the oxide (Parks, 1967). For example, silica has a lower ZPC than titania because Si^{4+} has greater polarizing power (high charge/radius ratio) than Ti^{4+}; the resulting higher acidity of Si-OH groups is manifested as a lower ZPC. In contradiction to this electrostatic model, Co^{2+} adsorbs at lower pH on TiO_2 than on SiO_2 (R.D. James and

Table 2. Predicted affinity sequences of divalent metals for oxides based on several metal properties

Property	Predicted order of affinity
Z^2/r	Ni > Mg > Cu > Co > Zn > Cd > Sr > Pb
K_1[a]	Cu > Pb > Ni > Co = Zn > Mg > Cd > Sr
Electronegativity (Pauling)	Cu > Ni > Co > Pb > Cd > Zn > Mg > Sr
Softness	Pb > Cd > Co > Cu > Ni > Zn > Sr > Mg
Irving–Williams series	Cu > Ni > Zn > Co > Mg > Sr

[a] K_1 is the first hydrolysis constant for the metal.
Source: Data from Huheey, 1972.

Healy, 1972a). "Specific" bonding forces must then be invoked to account for the behavior of Co^{2+} and transition metals in general.

A satisfactory model of transition and heavy metal adsorption must incorporate specific information about the particular metal-surface bonds formed. This is done in the constant capacitance models of oxides by defining explicitly the surface species formed in the chemisorption reactions and by allowing the equilibrium constants for the reactions (herewith) to quantify the affinity of metals for the surface (Schindler et al., 1976):

$$> S-OH \rightleftharpoons\ > S-O^- + H^+ (aq)$$

$$> S-O^- + M^{z+} \rightleftharpoons\ > S-O-M^{(z-1)+}$$

$$2 > (S-O^-) + M^{z+} \rightleftharpoons\ > (S-O)_2 = M^{(z-2)+}$$

Here, S is the metal ion of the oxide structure and M^{z+} is the adsorbed metal. The first reaction is clearly related to the oxide's ZPC, but the overall adsorption process will not show an obvious dependence on ZPC if the equilibria described by the second two reactions strongly favor adsorption.

The constant capacitance model assumes that all adsorption is by M^{z+}–H^+ exchange on surface OH groups and therefore that all metal-surface interactions occur via inner-sphere complexation (Goldberg, 1985). The fact that the stability of oxide-metal complexes has been correlated to the stability of the corresponding aqueous hydroxometal complexes (Kinniburgh et al., 1976; Schindler et al., 1976) is taken as evidence that the ligand properties of aqueous OH^- and $-S-OH$ groups are comparable. An alternative viewpoint has recently been proposed in which chemisorption and precipitation are seen as end-members of a sorption continuum (Farley et al., 1985). The fact that no clear sorption maximum is observed on oxides and slow adsorption follows the rapid initial adsorption seems to support this viewpoint (McLaren et al., 1981). Hydrolysis and sorption are then seen to be correlated because of the direct involvement of OH^- in the sorption of metal hydroxy species by oxides. This role of nucleation and precipitation reactions in the "sorption" of metals will be discussed further later in this chapter.

A complicating factor for models of metal adsorption on oxides is the formation of ternary complexes. Although some metal-complexing ligands such as ethylenediaminetetraacetate (EDTA) suppress metal adsorption (Bourg and Schindler, 1979; Elliot and Huang, 1979), others enhance adsorption by forming stable surface-metal-ligand (ternary) complexes (Davis and Leckie, 1978; Bourg et al., 1979; Elliot and Huang, 1981). These ligands are capable of complexing or chelating the metal ion while allowing simultaneous coordination of the metal to the surface. Bidentate ligands (e.g., ethylenediamine, bipyridyl, glycine) seem particularly effective in forming such complexes. One example of a ternary complex involving Cu^{2+} and glycine bonded at crystal steps of gibbsite is depicted in Figure 10. The metal/ligand ratio is critical in determining whether metal

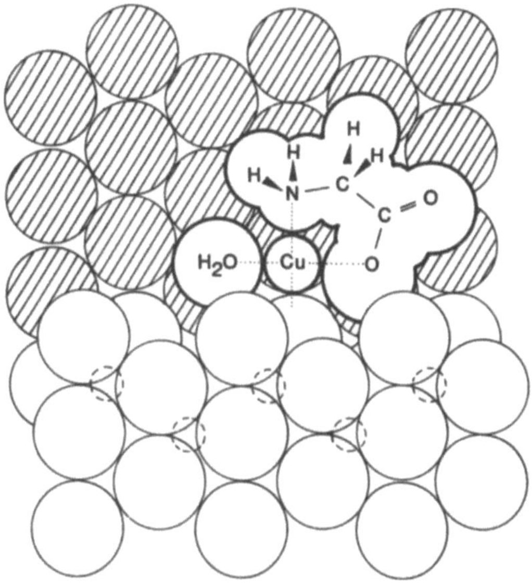

Figure 10. View down the *c*-axis of gibbsite of the suggested structure of Cu (glycine)$^+$ bound at a crystal step on the (001) surface.

adsorption at surfaces is enhanced or inhibited; a large excess of soluble complexing ligands obviously favors metal desorption relative to ternary complex formation (McBride, 1985a).

Inorganic as well as organic ligands can be involved in ternary complex formation. The fact that Zn^{2+} adsorption by iron and aluminum oxides, clays, and soil is enhanced by the presence of phosphate or sulfate (Stanton and Burger, 1970; Newton, 1971; Bolland et al., 1977; Moraes, 1982; Shuman, 1986) has led to the suggestion that Zn-phosphate or Zn-sulfate surface complexes form. A similar mechanism has been indicated for Cu^{2+} adsorption on allophane in the presence of phosphate (Clark and McBride, 1985) and Cd^{2+} adsorption on oxides in the presence of sulfate (Benjamin and Leckie, 1982). Nevertheless, ESR has produced equivocal results for the surface-bound metals, revealing only that Cu^{2+} coadsorbed with phosphate on allophane shows no evidence of inner-sphere coordination to phosphate (Clark and McBride, 1985), whereas VO^{2+} coadsorbed with phosphate on Al oxides and allophane forms an inner-sphere complex with phosphate (McBride, 1987). Phosphate adsorption by gibbsite is enhanced by the presence of metals such as Ca^{2+} and Cd^{2+}, again suggesting the adsorption of a metal-phosphate complex (Helyar et al., 1976). The structure of these ternary surface complexes is unclear, but present evidence suggests that the metal remains bonded to the oxide (Clark and McBride, 1985; McBride, 1985b). Possibly, a metal-phosphate ion pair forms in such a way that both metal and phosphate are bonded to the surface.

Some workers have claimed that the presence of metal cations does not enhance phosphate sorption on oxides unless a surface precipitate of the metal hydroxide forms (Benjamin, 1983). Certainly, the precipitation of metal phosphates or of metal hydroxides on which phosphate subsequently would adsorb may confuse the issue of ternary complex formation involving phosphate and other oxyanions. However, monomeric metal-phosphate pairs do appear to form at oxide surfaces (McBride, 1987), which suggests that metal-enhanced phosphate adsorption and phosphate-enhanced metal adsorption are dependent on this same reaction. A continuum probably exists between the initial formation of ion pairs and the surface precipitation of an identifiable metal-phosphate phase, so that the effects observed may depend very much on the metal/phosphate ratio and the loading level of these ions on the oxide surface sites.

The reversibility of sorption processes becomes a critical issue in any attempt to evaluate the efficiency of micronutrient additions to soils or the potential deleterious effects of heavy metal accumulations in soils. All evidence points to the conclusion that sorption of heavy metals on Fe and Al oxides is an inner-sphere complexation that does not obey the reversible mass-action relationships predicted for simple cation exchange. For example, Pb^{2+} adsorption on goethite, written as:

$$-\text{Fe-OH} + Pb^{2+} = -\text{Fe-O-}Pb^{+} + H^{+}$$

is insensitive to the concentration of $NaNO_3$ in the suspension (Hayes and Leckie, 1986). This implies that metals chemisorbed by inner-sphere complexation are unlikely to be very exchangeable by cations that have no specific affinity for the oxide. Adsorbed Cu^{2+} and Co^{2+} on allophanes, for example, are almost completely unexchangeable by Ca^{2+}, yet Pb^{2+} displaces most of the Cu^{2+} (Clark and McBride, 1984). Evidently, metals with high affinities for the bonding sites can readily displace preadsorbed metals. Similarly, H^{+} ions are usually able to displace a fraction of chemisorbed metals; that is, adsorption reactions such as the one (noted earlier) for Pb^{2+} are partially or completely reversible by pH change. This varies depending on the metal cation involved and the reaction time. In studies of various synthetic Mn oxides, McKenzie (1980) found that much of the adsorbed Cu^{2+} was acid extractable, whereas most of the Pb^{2+} was not, and extractability diminished after longer adsorption times. Murray (1975) noted that the more strongly adsorbed heavy metals on Mn oxides failed to desorb appreciably on lowering the pH. Part of the Cu^{2+} adsorbed on Al hydroxides, allophanes, and imogolite upon raising the pH is not rapidly desorbed by lowering the pH (McBride, unpublished data). On oxides such as goethite, the ability of strong acids to reverse adsorption of Ni^{2+}, Zn^{2+}, and Cd^{2+} diminished with time, a fact that has been attributed to slow diffusion into the solid phase (Gerth and Brümmer, 1981).

A significant fraction of sorbed Cu^{2+} on oxides of Fe and Mn is not isotopically exchangeable (McLaren and Crawford, 1974). Similarly, essentially all of the initially sorbed Co^{2+} on a mixed Fe-Mn "soil oxide" became isotopically nonexchangeable in a matter of a few weeks (McLaren et al., 1986). Since the sorbed ions that are not isotopically exchangeable are presumed not to be in equilibrium with the solution phase, the observed conversion of metals from a labile (isotopically exchangeable) to nonlabile form is an indication that sorption involves processes that are not wholly reversible. The process of isotope exchange:

$$-S-O-M + M^* \rightleftharpoons -S-O-M^* + M$$

requires the breaking of a metal-oxide bond and therefore may be relatively slow because of an activation energy that has to be overcome, but as the bond formed is obviously no stronger than the bond broken, self-exchange may be less efficient than exchange by a preferred metal in which the free energy of the surface is changed.

Different measures of reversibility or lability are used in assessing metal availability. These include (1) exchangeability by cations that do not specifically adsorb, (2) exchangeability by specifically adsorbing cations, (3) pH reversibility, (4) isotope exchangeability, (5) desorbability by chelating agents, and (6) dissolution by strongly acidic solutions. Yet, assuming that adsorption is achieved by increasing the metal ion concentration, the reversibility of the adsorption reaction is properly measured in terms of the desorption induced in response to a reduction of the solution concentration of the metal. Generally, when this is done, very little desorption of heavy metals occurs (McLaren et al., 1986), but whether the surface reaction is genuinely irreversible or simply very slow in the reverse direction is unclear. Studies of Pb^{2+} adsorption and desorption by goethite suspensions have indicated that the adsorption reaction step is fast and probably diffusion controlled ($K_1 \approx 2 \times 10^5$ mol^{-1} dm^3 s^{-1}), whereas the desorption is much slower ($K_{-1} \approx 6 \times 10^2$ mol^{-1} dm^3 s^{-1}), probably limited by the activation energy required to break the Pb^{2+} surface bond (Hayes and Leckie, 1986). According to chemisorption theory, adsorption may or may not require a significant activation energy, E_a^*, but desorption always requires an activation energy, E_d^*, since desorption necessitates that the energy of adsorption, ΔH, be overcome (Adamson, 1976). Thus, as illustrated by Figure 11, E_d^* is given by:

$$E_d^* = E_a^* + \Delta H$$

The result is a much higher activation energy for desorption than for adsorption and an adsorption rate that is much faster than the desorption rate. In addition, since the energy of adsorption, ΔH, is usually higher at low levels of chemisorption, the rate of desorption will be even slower at low sorption levels. Thus, the commonly reported hysteresis in metal sorption reactions may actually reflect a nonequilibrium condition caused by slow desorption rather than true irreversibility. The explanation of hysteresis cannot be fully based on slow reaction rates, however, as experiments have proven that part of the metals slowly become occluded in iron oxides (Gerth and Brümmer, 1983) and can be released to solution only by dissolution of the oxide. The evidence points to solid diffusion and the formation of solid solutions, yet it is generally believed that solid-state diffusion is insignificant at room temperature (Driessens, 1986). A better understanding of these slow processes will require modern techniques of surface analysis and

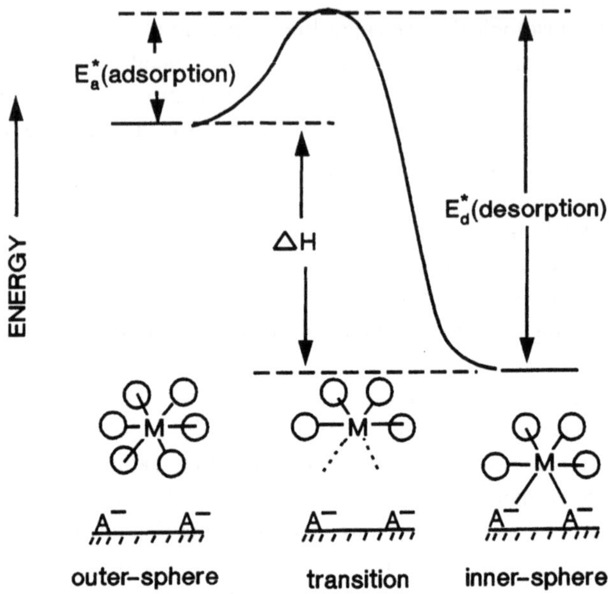

Figure 11. Energy diagram for the adsorption of a metal, M, at anion surface sites, A^-.

is obviously a subject that needs further study and clarification. Further discussion of precipitation and solid-solution formation, one potential mechanism operative over long time periods, appears later in this chapter.

The operational definitions of reversibility should not be confused with the strict thermodynamic definition of reversible processes. In principle, one could reversibly adsorb or desorb metals at oxide surfaces by making small changes in metal concentration, so that an equilibrium pathway between the initial and final states was followed. In reality, most adsorption or desorption experiments are conducted under a strong driving force for practical reasons, that is, the intermediate states of the process are not equilibrium states and no unique pathway connects the initial and final state. We conclude by this definition that all real processes are more or less irreversible. Nevertheless, whether the metal adsorption reaction proceeds under a strong driving force or by a reversible pathway does not affect the energy, ΔE, of the reaction, since ΔE depends only on the initial and final states.

IV. Nucleation, Precipitation, and Solid Solutions

An underlying problem associated with adsorption experiments is distinguishing true chemisorption processes from precipitation. Means of making this distinction are sometimes available, but more often, the formation of a new solid phase during the sorption process is not recognized. Mineral surfaces catalyze the nucleation of crystals; this process of heterogenous nucleation is considered to be the most important mechanism of crystal formation in natural waters (Stumm and Morgan, 1981). The mineral reduces the energy barrier for the nuclei of crystals to form from solution by providing a sterically similar, yet chemically foreign, surface for nucleation. The energy barrier arises from the fact that the small crystallites that must initially form in the crystallization process are more soluble than large crystals because of a higher interfacial energy between small crystals and solution. Thus, homogeneous nucleation and precipitation (in the absence of a foreign surface) cannot occur unless the solubility product has been sufficiently exceeded to form crystal nuclei. It is well known, for example, that solutions supersaturated with respect to gibbsite do not always precipitate gibbsite (Hsu, 1977). The presence of a surface reduces the extent of supersaturation necessary for precipitation to an extent determined by the similarity of lattice dimensions of the surface and the precipitating solid. Smectite can promote gibbsite crystallization in partially hydrolyzed Al solution, for example (Hsu, 1977). In the case where the "seed" crystals are structurally identical to the crystals being formed, no supersaturation is necessary for precipitation to occur.

There is general consensus that transition and heavy metal solubility in soils, with the possible exceptions of Fe and Mn, is *not* controlled by the solubility product of a pure solid phase (Brümmer et al., 1983; Herms and Brümmer, 1984). One obvious reason for this is that at low concentrations of most metals in soil solids, adsorption processes are able to maintain solubility at a level too low for precipitation to occur. For example, small quantities of Zn^{2+}, Mn^{2+}, or Cd^{2+} added to aqueous suspensions of $CaCO_3$ are sorbed on the carbonate surface, reducing the concentration of the metal below that predicted from the solubility of the pure metal carbonates (Jurinak and Bauer, 1956; McBride, 1979a, 1980a; Brümmer et al., 1983). Nevertheless, there are cases where precipitated solid phases may determine metal solubility, usually situations of high metal loadings. Calcareous soils contaminated with Cd^{2+} show evidence of being saturated with respect to $CdCO_3$ precipitation (Cavallaro and McBride, 1978). Pb^{2+} and Zn^{2+} solubility in alkaline soils may be controlled by phosphate and silicate solid phases, respectively (Herms and Brümmer, 1984; Brümmer et al., 1983). In wet soils, Zn^{2+} solubility is potentially subject to control by ZnS formation (Gilmour and Kittrick, 1979). Malachite ($Cu_2CO_3(OH)_2$) has been identified as a deposit on carbonate rocks of a Cu-contaminated soil in the field (McBride and Bouldin, 1984). These examples involve predominantly high pH soils. Solubility data for acid mineral and organic soils indicate that these soils remain highly undersaturated with respect to precipitation of metals as hydroxides or carbonates, even in cases where high levels of metals have been added to the soils (McBride and Blasiak, 1979). Plots of pH versus log (M^{2+}), where (M^{2+}) is the concentration of soluble divalent metal in soil suspensions, have slopes between 1.0 and 2.0 rather than the slope of 2.0 expected for precipitation of pure hydroxides or carbonates. In Figure 12, typical Cu^{2+} solubility lines are plotted, showing that less acid soils have slopes closer to 2.0. As increasing quantities of metal are added to the soil, the solubility line more closely approaches the line of a pure

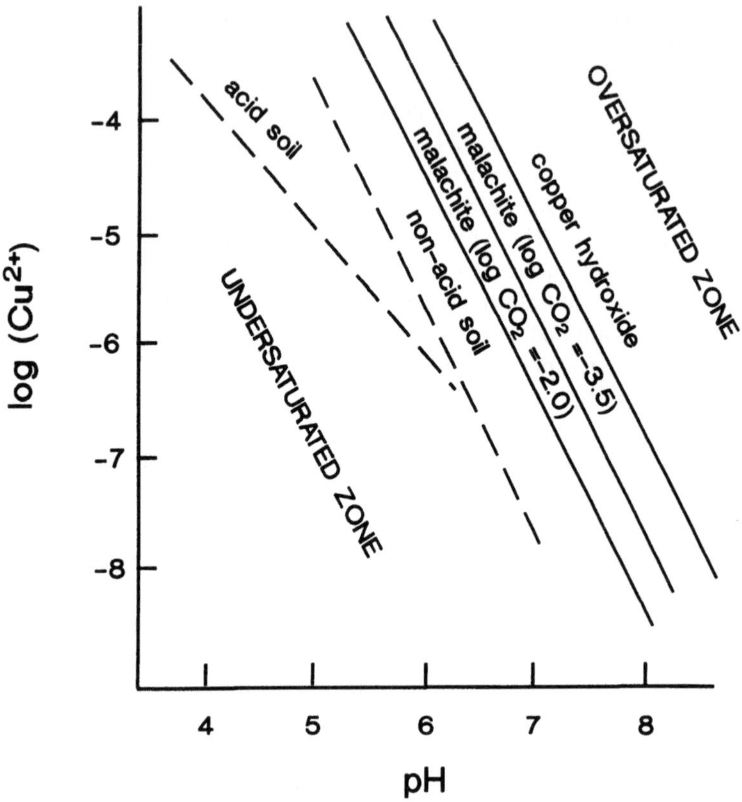

Figure 12. Typical relationships of Cu^{2+} activity to adjusted soil pH for acid and calcareous soils. Solubility lines for precipitated Cu phases are shown for reference. (Based on data from Cavallaro and McBride, 1978.)

mineral precipitate (McBride and Blasiak, 1979). This is demonstrated in Figure 13 for Zn^{2+} solubility at four levels of Zn added to a mineral soil.

Failure of the "solubility product" approach, even in cases where high metal levels in soils might be expected to produce precipitates, could be attributed to one of a number of factors. First, it is not always possible for the free metal ion concentration to be experimentally measured separately from metal-ligand complexes in order to obtain an accurate value of the ion activity product. Total soluble metal concentration cannot in general be assumed to approximate the free metal ion concentration (as was implied in the solubility plots of Figures 12 and 13). This issue will be further elaborated later in a discussion of metal speciation. Second, the assumption that equilibrium has been reestablished after the soils are moved from the field into the laboratory may not be justified. There is serious doubt whether soils in the field ever attain equilibrium beyond the microscopic scale (Chesworth and Dejou, 1980). Although one might expect that vigorous mixing of soils suspended in aqueous solution in the laboratory would overcome

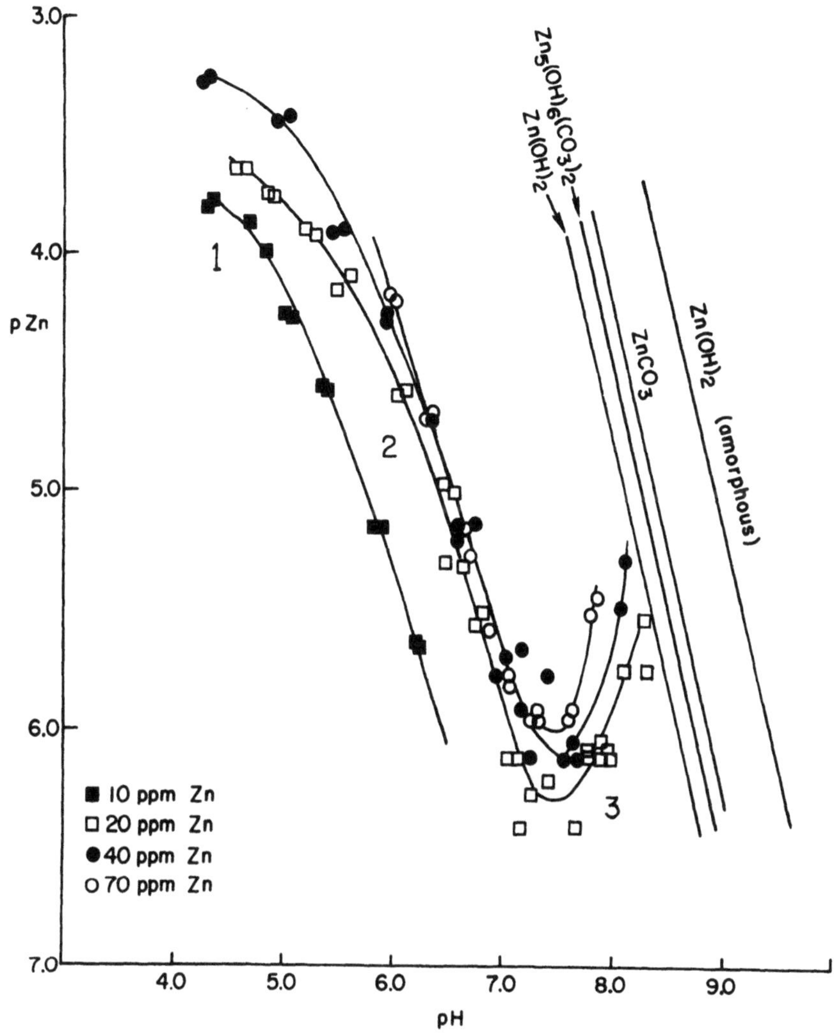

Figure 13. Relationship between Zn^{2+} concentration in soil solution and pH at four levels of Zn added to an acid mineral soil. Solubility lines for precipitated Zn phases are shown for reference, assuming atmospheric levels of CO_2.

this difficulty, the soil solids are likely to undergo a sluggish response to the changed O_2 and CO_2 concentrations, pH, and electrolyte concentration. In particular, a drastic change in the CO_2 level is predicted to significantly alter the solubility of many metals. Concentrations of CO_2 in soil air can be higher than 1% (Fernandez and Kosian, 1987). Figure 12 shows that a decrease of CO_2 from field concentration to laboratory concentration (350 ppm) could raise the solubility of Cu^{2+} by a factor of greater than 5 if malachite were the controlling solid phase. The figure also suggests, however, that even very high levels of CO_2 in soils could not sufficiently lower the solubility of Cu carbonate minerals to account for the very low concentrations of free Cu^{2+} in soil solutions.

The "solubility product" approach requires that the least soluble mineral able to crystallize from soil solution be identified. In the case of Cu, Zn, and other metals, it has been suggested that mixed oxide minerals such as inverse ferrites, which are generally less soluble than simple metal oxides, may maintain low metal solubilities (Lindsay, 1979). The observation that Zn and Cu solubility diminishes in flooded soils as soluble Fe and Mn increase has been proposed to indicate ferrite formation (Iu et al., 1981a, 1981b; Sajwan and Lindsay, 1986). Although the raised pH resulting from chemical reduction processes in acid soils could in itself be responsible for part of this effect, additional processes appear to be lowering the extractability of these metals. Suggested explanations include the formation of a fresh Fe or Mn oxide surface in intermittently waterlogged soils that have high adsorptive capacities for the metals or the coprecipitation of Fe with metals to form inverse ferrites such as $ZnFe_2O_4$ or $CuFe_2O_4$ (Lindsay, 1979). It remains to be proven that these mixed oxides, which have Fe^{3+} in tetrahedral coordination, will form at the temperature, pressure and pH conditions typically found in soils. The formation of $MnFe_2O_4$, for example, appears to require high pH and high temperature (Cornell and Giovanoli, 1987).

Coprecipitation of metals as impurities in Al, Fe, and Mn oxides may confound the "solubility product" approach by producing solid solutions with ill-defined stoichiometry. There is considerable evidence for the coprecipitation of metals from soil solutions, but the nature of the precipitate is usually unknown. In some soils, for example, the reduction of diethylenetriaminepentaacetic acid (DTPA)-extractable Fe induced by liming acid soils was correlated with a reduction in extractable Zn (Jahiruddin et al., 1986). Also, additions of Zn to soils have tended to cause reductions in soluble Fe^{3+} (Pulford, 1986). Although one might be tempted to hypothesize Zn-ferrite precipitation in these cases, the results should be analyzed cautiously because soluble organic matter can play a role in the "coprecipitation" of metals. Acidification of soil solution, which lowers the solubility of organic matter, may decrease the concentrations of soluble Al^{3+} and Fe^{3+} (B.R. James and Riha, 1984, 1986; M.G. Johnson, 1986), presumably by removing the organically complexed metals from solution. A similar process may precipitate heavy metals from solution. The data of Pulford (1986), revealing an inverse relationship between soluble Zn and soluble Fe, is probably related to the tendency of excess Zn^{2+} to reduce the solubility of organic matter and precipitate Fe^{3+}-organic complexes. The fact that detectable concentrations of total soluble Fe were measured at all is an indication that either $Fe^{3\pm}$ organic complexes or Fe^{2+} were predominant species. As Ca^{2+} addition to the soils eliminated soluble Fe, the only reasonable interpretation of these data is that organic ligands retained Fe^{3+} in soluble form until metal ions were added, which then caused flocculation and precipitation of the organics.

The apparent coprecipitation of metals such as Zn with Fe and Al during the liming of soil can reduce the solubility of these metals, but careful analysis suggests that Zn is largely adsorbed on the surfaces of the freshly precipitated oxides rather than incorporated into these solids (Jahiruddin et al., 1986). The observations that Cu^{2+} coprecipitated in noncrystalline $Al(OH)_3$ is predominantly in the very fine particles (McBride, 1978a) and that transition metals (e.g., Cu^{2+}, Mn^{2+}, Co^{2+}, Ni^{2+}) inhibit oxide crystallization (Nalovic et al., 1975) point to the conclusion that metals are concentrated at surfaces. In fact, ESR studies have shown little or no difference in the chemical environment of Cu^{2+} and VO^{2+} whether these metals were coprecipitated with $Al(OH)_3$ or adsorbed on preexisting $Al(OH)_3$ surfaces (McBride, 1978a). Much of the "coprecipitated" Cu^{2+} was accessible to react with complexing or reducing agents (McBride, 1978a, 1982a). Thus, although rapid precipitation of Al and Fe hydroxides might "occlude" M^{2+} ions, the metals appear to be distributed near the surface of the solid phase. Although Mn substitution in goethite and hematite has been reported (Cornell and Giovanoli, 1987), it is limited to less than 15 and 5 mole %, respectively, and there is evidence that only the Mn(III) oxidation state can be comfortably accommodated in the goethite structure (Stiers and Schwertmann, 1985). The hematite structure must be even less adaptable to substitution than goethite because Mn substitution stabilizes goethite relative to hematite (Ebinger and Schulze, 1986). It is unclear how the structural charge from Mn^{2+} substitution into Fe(III) hydroxides could be internally compensated; perhaps structural OH^- groups are protonated to maintain a neutral structure. Otherwise, substitution would be restricted to the surface, where charge can be easily compensated, or to M^{3+} ions that do not generate structural charge. In fact, metal ions such as Cr^{3+}, V^{3+}, and Al^{3+} are known to isomorphously substitute into Fe oxides (Nalovic et al., 1975; Schwertmann et al. 1977; Murad and Schwertmann, 1983), but the foreign ions cause a lower degree of crystallinity even in these cases.

Given these observations, the hypothesis of true solid diffusion to account for reduced extractability of heavy metals adsorbed on oxides is difficult to support; in most cases, the structure would be destabilized by the penetration of divalent metals below the surface layer of the oxide. A more attractive hypothesis may be very slow diffusion of metals into extremely small pores of particle aggregates, a process that is likely to display a high degree of apparent nonreversibility.

An alternative hypothesis, one that involves the formation of solid solutions at the oxide surface, should also be considered at this point. If solid solutions form, incorporating metal ions from both solution and the oxide surface, they are likely to be of highly variable composition and stability. However, some well-defined mixed hydroxide compounds are known. For example, hydrotalcite-like compounds that consist of positively charged brucite-like layers, $[Mg_{1-x}Al_x(OH)_2]^{x+}$, alternating with interlayers of exchangeable anions can form if $0 < x < 0.33$ (Mortland and Gastuche, 1962; Miyata, 1983), and they appear to be particularly stable for $x = 0.3$. These are readily synthesized by coprecipitating salts of Al^{3+} and Mg^{2+} at high pH, and numerous analogous structures that incorporate metals such as Ni^{2+} and Zn^{2+} in the brucite layers are possible (Miyata, 1975). Although the conditions for synthesis suggest that hydrotalcites could form only in highly alkaline, saline environments, positively charged brucite-like layers analogous to these structures readily form in the interlayer regions of layer silicates. The negative charge of the silicate sheet acts as the charge-compensating anion, thus stabilizing the hydrotalcite-like layer. Heavy metals of appropriate charge and radius may be able to act as a proxy for Mg^{2+} in the interlayer. Evidence for this has been suggested by the strong selectivity of smectites interlayered with Al hydroxy polymers for Cu^{2+} and Zn^{2+} relative to Pb^{2+} and Cd^{2+} (Keizer and Bruggenwert, 1986). The latter metals are too large to be accommodated in the octahedral coordination sites of hydrotalcite. The selectivity cannot be explained by chemisorption on the surface of the hydroxy-Al polymers as chemisorption invariably favors Pb^{2+} over Zn^{2+} (Kinniburgh et al., 1976). Thus, the formation of mixed hydroxide interlayers may be an active mechanism in soil, converting heavy metals into nonavailable forms.

The theory of solid solutions predicts that the solubility of a heavy metal can be lowered in a mixed ionic compound relative to the solubility of the pure compound. Our discussion summarizes this theory as outlined by Driessens (1986). Consider, for example, a heavy metal cation, B, isomorphously substituted into a solid composed of metal cations, A, and anions, Y. The chemical formula, $A_{1-x}B_xY$, is variable because x can range from 0.0 to 1.0 if AY and BY form a continuous solid-solution series. Unlike ionic compounds of fixed composition, solid solutions do not have constant solubility products; rather, equations for both components, AY and BY, must be specified:

$$(a_A)\,(a_Y) = K_{sp}{}^{AY}\, a_{AY,s} \tag{5}$$

$$(a_B)\,(a_Y) = K_{sp}{}^{BY}\, a_{BY,s} \tag{6}$$

Here, a_A, a_B, and a_Y represent the activities in aqueous solution of the ions A, B, and Y, $K_{sp}{}^{AY}$ and $K_{sp}{}^{BY}$ are the solubility products of *pure* AY and BY, whereas $a_{AY,s}$ and $a_{BY,s}$ are the activities of the components AY and BY in the solid solution. At equilibrium, the distribution of A and B ions between the aqueous (aq) and solid phase (s) is represented by the reaction:

$$AY^{(s)} + B^{(aq)} \rightleftharpoons A^{(aq)} + BY^{(s)}$$

If the solid phase is homogeneous, the reaction can be described by:

$$\left(\frac{a_{BY,s}}{a_{AY,s}}\right) = D\left(\frac{a_B}{a_A}\right) \tag{7}$$

where D is the distribution coefficient that quantifies the relative extent to which the two metal cations are incorporated into the solid. For an ideal solution of formula $A_{1-x}B_xY$, the heat of mixing is zero; that is, one cation can substitute for the other without any change in the energy of the solid structure. In addition, the theoretical entropy of mixing in this ideal solid solution is given by:

$$S = 2.303R[x \log x + (1 - x) \log (1 - x)] \tag{8}$$

This means that for the ideal case, the activities of the components are given simply by:

$$a_{AY,s} = 1 - x \tag{9}$$

$$a_{BY,s} = x$$

If the activity coefficients of ions A and B in solution are essentially the same, equation (7) can be rewritten in terms of concentrations:

$$\frac{[BY]_s}{[AY]_s} = D\frac{[B]_{aq}}{[A]_{aq}} \tag{10}$$

where the subscripts s and aq denote mole fractions in the solid and mole concentrations in the solution, respectively. Even for this ideal case, where D is expected to be constant, a value of D that is very different from unity will mean that a solid solution precipitated from soil solution cannot be homogeneous because one metal will be selectively scavenged from solution as the solid crystallizes and the less preferred ion will concentrate in the outer layer of the crystal. The significance of this mechanism in soils can easily be imagined when a heavy metal is buried within a crystal and becomes unavailable for subsequent desorption. The solubility of a trace metal, B, can be lowered to a level well below that predicted from the solubility product of the pure solid phase, BY. Combining equations (6) and (9) produces the relationship

$$(a_B)\,(a_Y) = x\,K_{sp}{}^{BY} \tag{11}$$

This reveals that very low levels of substitution of B into the structure could produce an "effective" solubility product, $(a_B)\,(a_Y)$, that is orders of magnitude below the K_{sp} of the pure BY solid.

Studies of Cd^{2+} and Mn^{2+} removal from solution in the presence of calcite have indicated an initial fast reaction followed by a slow process (McBride, 1979a, 1980a; Davis et al., 1987). For Cd^{2+}, the fast process appears to be exchange of Ca^{2+} at the hydrated $CaCO_3$ surface:

$$Cd^{2+} + CaCO_3 \rightleftharpoons CdCO_3 + Ca^{2+}$$

This reaction becomes more prevalent as the quantity of $CaCO_3$ surface in suspension is increased and proceeds even if the suspension is undersaturated with respect to solid $CdCO_3$ (Davis et al., 1987). Thus, the fast reaction is an adsorption rather than a precipitation process, although Cd^{2+} may diffuse into the hydrated surface layer with time. The subsequent slow removal of Cd^{2+} from solution has been attributed to recrystallization of Cd^{2+} and Ca^{2+} in the poorly ordered surface layer to form a crystalline solid solution at the surface. This sequence of reactions explains the fact that the EDTA-extractable and isotopically exchangeable fraction of Cd in the solid phase was initially quite high but diminished with time (Davis et al., 1987). Reorganization at the surface into a solid solution requires that a significant rate of calcite dissolution and recrystallization occur, and this has been confirmed by Ca^{45} isotope exchange experiments (Davis et al., 1987). For an ideal $Ca_{1-x}Cd_xCO_3$ solid solution, the distribution coefficient as defined herein can readily be shown to equal the quotient of solubility products of the pure carbonate solids:

$$D = \left(\frac{K_{sp}{}^{CaCO_3}}{K_{sp}{}^{CdCO_3}}\right) = \left(\frac{10^{-8.47}}{10^{-11.3}}\right) = 680 \tag{12}$$

Clearly, a very strong tendency should exist for the recrystallizing calcite to incorporate Cd^{2+}. Equation (11) reveals that as long as the mole fraction of Cd^{2+} incorporated into the crystal is small, Cd^{2+} can be removed from solution despite a large degree of undersaturation with respect to pure $CdCO_3$ precipitation. The large value of D would cause initial recrystallization to incorporate higher concentrations of Cd^{2+} than later recrystallization. In any event, heavy metal occlusion into a foreign matrix seems to be a viable mechanism.

Even in cases where solid solutions play a role in controlling heavy metal solubility, it is unlikely that the solid solution is ideal. Most solid solutions are "regular," meaning that the heat of mixing of the components, AY and BY, is not zero because of an interaction energy, W, between ions A and B in the crystal. The heat of mixing, H_m, becomes (Driessens, 1986):

$$H_m = x(1 - x)W \tag{13}$$

The activities of the components for $A_{1-x}B_xY$ are then more complex functions of x than in the ideal case:

$$a_{AY,s} = (1 - x) \exp \{x^2 W/2.303 RT\} \tag{14}$$

$$a_{BY,s} = x \exp \{(1 - x)^2 W/2.303 RT\} \tag{15}$$

Equation (10) then becomes:

$$\left(\frac{[BX]_s}{[AX]_x}\right) = D \exp \{-(1 - 2x) W/(2.303 RT\} \left(\frac{[B]_{aq}}{[A]_{aq}}\right) \tag{16}$$

If equation (10) rather than equation (16) is applied to the usual case of regular solid solutions, D will appear to be variable, dependent on x. However, when the nonideal interaction term, W is accounted for (equation 16), D becomes constant. A negative value of W stabilizes the formation of the solid solution relative to the pure ionic solids, AY and BY, which are its components. In equation (16), this fact is reflected in the exponential term, which may be much greater than unity if the interaction term is negative and large and if the degree of substitution of B into AY is small ($x \ll 1$). The apparent value of D, as defined by equation (10), might then appear to be very large at low values of x because of stabilizing interactions. Conversely, positive interaction terms between A and B can cause the solid phase, AY, to discriminate against the inclusion of B.

In soils, the relative importance of solid-solution formation in controlling trace metal availability is difficult to evaluate. Particular minerals such as calcite are known to readily incorporate divalent metals of ionic radius equal to, or less than that of Ca^{2+}(e.g. Mn^{2+}, Zn^{2+}, Cd^{2+}, Fe^{2+}, Co^{2+}) (Pingitore, 1986), and soil calcites are substituted by Mn^{2+} and presumably various other metals (McBride, 1979a). However, each of the previously mentioned trace metals forms a carbonate that is isostructural with calcite, so that a high degree of miscibility of these metals in calcite is expected. In contrast, as discussed earlier, metals such as Cu^{2+}, Zn^{2+}, Ni^{2+}, and Co^{2+} appear to be excluded from hematite (α-Fe_2O_3) during its crystallization and concentrate at or near the surface of the oxide (Sidhu et al., 1980). Other oxide minerals such as magnetite ($Fe_2^{3+}Fe^{2+}O_4$) readily accept these trace metals and others in their structure, probably because isomorphous substitution of divalent transition metals into the sublattice sites normally occupied by Fe^{2+} is energetically favorable. Small trivalent metal ions such as Cr^{3+} and Mn^{3+} are expected to readily substitute for Fe^{3+} and Al^{3+} in oxide structures, and evidence suggests that this occurs (Sidhu et al., 1980; Stiers and Schwertmann, 1985; Cornell and Giovanoli, 1987). In addition, Ca-phosphate minerals in soils can be expected to form solid solutions with Pb^{2+} and Cd^{2+}, ions of similar size to Ca^{2+}. Hydroxyapatite in particular has an extremely strong tendency to incorporate Pb^{2+} into its structure (Driessens, 1986).

These examples make it apparent that size and charge of the metal ion determine the favorability of solid-solution formation. Whether solid solutions actually form in soils may be controlled more by slow rates of mineral dissolution – a necessary preliminary step for the recrystallization of solid solutions – than by thermodynamic favorability. Nevertheless, certain features of metal sorption in soils are consistent with (but do not prove) solid-solution formation. These include decreasing reversibility of adsorption with time, high selectivity for certain metals, and solubilities below those predicted from solubility products of the pure metal-containing solid phases.

The alternative viewpoint in rationalizing the loss in lability of heavy metals during slow sorption processes is that the metals diffuse into the crystalline structure of the absorbent. This is generally believed to be untenable, in that solid diffusion is extremely slow at ambient temperature. Davis et al. (187) have pointed out, for example, that Cd^{2+} diffusion beneath the noncrystalline hydrated surface layer of calcite is unlikely. On the other hand, assimilation of heavy metals into solid phases by recrystallization requires a *relatively* high-solubility product of the solid (as is the case for $CaCO_3$). In cases where heavy metals have been observed to become less available after sorption on Fe oxides (Gerth and Brümmer, 1983), the low solubility of the oxide would seem to rule out significant recrystallization within the experimental time scale. Dissolution of Fe and Mn oxides is greatly enhanced by chemical reduction, so that alternate reduction and oxidation of soils might be expected to improve the chances for trace metal occlusion by oxides. Some evidence exists to support this hypothesis, although oxides freshly precipitated from Mn^{2+} or Fe^{2+} solutions may simply be effective adsorbents for trace metals (Iu et al., 1981b). In the absence of chemical reduction, the only other reasonable explanations of a gradual loss in reversibility of heavy metal "adsorption" on oxides is diffusion into thick hydrated surface layers of poorly crystalline oxides or diffusion into very narrow intercrystal discontinuities or pores. Further research is evidently needed on this important question of reversibility.

V. Redox Processes Affecting Metal Solubility

Biological activity can greatly alter the solubility of metals in soils by causing, directly or indirectly, changes in the oxidation state of metals. However, even considering the long-studied case of Mn^{2+} oxidation in soils and natural waters, the importance of microbial relative to chemical "catalysis" of oxidation is difficult to assess (Davies, 1986). There is an unfortunate tendency to assume, in the absence of proof to the contrary, that soil-mediated redox processes are biological. Some of the known nonbiological electron-transfer processes that can affect metal solubility in soils are now described. Their importance relative to microbial processes will depend on the specific chemical and mineralogical properties of the soil.

A. Oxidation of Metals by Metal Oxides

Generally, heavy metals are less soluble in their higher oxidation states. Therefore, the ability of Mn oxides (and to a lesser extent Fe oxides) to directly oxidize metals or to catalyze metal oxidation by O_2 could provide a mechanism for lowering trace metal solubility. Several examples have been reported in the literature where Mn oxides reduce the solubility of metals by an oxidation process. These include the oxidation of adsorbed Co(II) to Co(III) (Dillard and Schenck, 1986) and Fe(II) to Fe(III) (Stumm and Morgan, 1981). Oxidation of Mn^{2+} is autocatalytic (Stumm and Morgan, 1981), so that Mn oxides show large apparent sorption capacities for Mn^{2+} that increase with pH (J.W. Murray, 1975). In contrast, Mn^{2+} oxidation in aqueous solution is very slow below pH 8. Therefore, in the absence of Mn-oxidizing organisms, Mn oxides could have an important role in reducing the concentrations of Mn^{2+} in soil solutions.

Other metals such as Pb^{2+} have been suggested to oxidize to insoluble forms at Mn oxide surfaces (Hem, 1978), although McKenzie (1980) found no evidence for Pb^{2+} oxidation by Mn oxides, despite the remarkable sorption capacity of the oxides for this metal. It is difficult to distinguish the processes of metal chemisorption, coprecipitation, and electron transfer on Mn oxides as all have the effect of strongly scavenging heavy metals from solution. Changes in the average oxidation state of the oxide or the retained metal are not often confirmed in Mn-oxide sorption experiments because of the lack of available techniques that provide direct information on metal oxidation states. The release of Mn^{2+} into solution upon heavy metal adsorption is usually assumed to indicate a redox process, but it can also arise from the exchange of Mn^{2+} from surface sites (Traina and Doner, 1985b).

One example where electron transfer between the adsorbed metal and Mn oxide may *increase* metal solubility is Cr(III) oxidation to Cr(VI). Oxidized Mn in soil has been correlated to the tendency of soils to oxidize Cr^{3+} to chromate (Bartlett and James, 1979). In natural aerated waters, the oxidation of Cr(III) to Cr(VI) by O_2, which is predicted to be a favorable process thermodynamically, is greatly accelerated in the presence of Mn^{2+} (Osaki et al., 1980). Oxidation of

Cr^{3+} in soil and water may have the same chemical mechanism, which involves the autooxidation of Mn followed by catalysis of Cr oxidation. Because the chromate anion is generally more soluble than Cr^{3+} in soils and much more toxic to animals, oxidation increases the environmental hazard of this element.

The Fe oxides are less powerful oxidizing agents than Mn oxides, and there is little evidence that they directly oxidize metals. Nevertheless, they may play a role in catalyzing oxidation by O_2. For example, Mn^{2+} oxidation by O_2 above pH 7.0 is more rapid in the presence of Fe oxides than in homogeneous solution (Davies, 1986). A mechanism involving the initial formation of a Mn^{2+}-surface complex:

$$\begin{array}{l}\text{-Fe-OH}\\ \\ \text{-Fe-OH}\end{array} + Mn^{2+} \rightleftharpoons \left.\begin{array}{l}\text{-Fe-O}\\ \\ \text{-Fe-O}\end{array}\right\rangle \text{Mn(II)} + 2H^{+}$$

followed by the adsorption of O_2 and subsequent electron transfer to produce a Mn(III)-superoxide complex:

$$\left.\begin{array}{l}\text{-Fe-O}\\ \\ \text{-Fe-O}\end{array}\right\rangle \text{Mn(II)} + O_2 \rightleftharpoons \left.\begin{array}{l}\text{-Fe-O}\\ \\ \text{-Fe-O}\end{array}\right\rangle \text{Mn(II)}-O_2$$

$$\left.\begin{array}{l}\text{-Fe-O}\\ \\ \text{-Fe-O}\end{array}\right\rangle \text{Mn(II)}-O_2 \rightleftharpoons \left.\begin{array}{l}\text{-Fe-O}\\ \\ \text{-Fe-O}\end{array}\right\rangle \text{Mn(III)}-O_2^{-}$$

has been proposed to explain catalysis (Davies, 1986). Cations such as Mg^{2+}, which can compete with Mn^{2+} for sorption sites on oxides, or anions such as salicylate, which may enter into ligand exchange reactions with surface Fe-OH groups, inhibit the catalytic process (Davies, 1986). Also, pH is critical to this mechanism since Mn^{2+} adsorption on goethite, for example, occurs above pH 7 (Bleam and McBride, 1985). Therefore, catalysis of Mn^{2+} oxidation by Fe oxides is likely to be significantly only at soil pH values near or above 7.

B. Dissolution of Metals by Organics

Organic molecules with the capability to complex with metals can potentially increase concentrations of these metals in soil solution by dissolution reactions at mineral surfaces (Jorgensen, 1976; Manley and Evans, 1986; Pohlman and McColl, 1986). Generally, the metal-complexing ability of the organic within the range of soil pH is a good indicator of its ability to adsorb to metal oxides and enhance mineral dissolution. Dissolution can be assisted or enabled by redox reactions in which the organic reduces the metal ion at the surface. In the case of Fe oxides, the general reaction is:

$$\text{Fe(III)}-\text{organic complex} \rightarrow \text{Fe(II)} + \text{oxidized organic}$$

This reaction is promoted at low pH and is facilitated for Fe(III)-organic complexes involving phenolic groups (Hider et al., 1981). It is interesting that soil fulvic acids reduce Fe(III) by an apparently similar process (Goodman and Cheshire, 1987). The reaction is promoted by UV light to varying degrees, depending on the nature of the organic ligand. Metal-organic complexes with strong ligand-to-metal charge-transfer transitions in the visible and UV portion of the spectrum show photochemical activity that can rapidly dissolve the iron oxide and oxidize the adsorbed organic (Waite, 1986). Hydroxycarboxylate complexes with Fe(III) are particularly photochemically active, with the result that molecules such as salicylate induce Fe^{2+} release into solution. Oxalate has a well-known ability to generate Fe^{2+} in the presence of light and Fe(III) oxides, and this type of reaction is probably important in the dissolution of iron oxides in natural waters (Zinder et al., 1986). Fulvic acids also possess some photochemical activity (Waite and Morel, 1984) in addition to their known ability to increase metal solubility by forming soluble complexes.

If molecular oxygen is present in soil solution, then the redox reaction is coupled to the reoxidation of Fe^{2+}:

$$\mathrm{Fe(II)} + \tfrac{1}{4}\mathrm{O_2} + \text{organic} \rightarrow \mathrm{Fe(III)}-\text{organic complex}$$

and the Fe(III) is effectively recycled, acting as a "catalyst" for the oxidation of organics by O_2 (Stumm and Morgan, 1981). However, if anaerobic conditions exist in soil solution, high concentrations of Fe^{2+} may accumulate.

A wide range of organic compounds can be oxidized at Mn-oxide surfaces. These include diphenolics (substituted and unsubstituted catechols, hydroquinones, and resorcinols), salicylic acid, pyruvic acid, oxalic acid, and fulvic acids (Stone and Morgan, 1984a, 1984b). The redox reaction with fulvic acid is photocatalyzed, presumably by a similar charge-transfer mechanism to that proposed for synthetic Fe oxide-organic complexes. Some organics, such as malate (Jauregui and Reisenauer, 1982) and glutamate (Traina and Doner, 1985a), are oxidatively decomposed by Mn oxides. It is likely that reactions of this type occur in soils, particularly in the vicinity of roots that exude low-molecular-weight organic acids. Certain monophenols—particularly those that have electron-donating substituents on the aromatic ring—when added to soils containing Fe and Mn oxides are chemically oxidized, releasing Fe^{2+} and Mn^{2+} as soluble products (Lehmann et al., 1987). Hydroquinone is commonly used to extract "easily reducible" Mn from soils, and this reaction is very rapid in soil (McBride, unpublished data), an indication that the redox process is a chemical rather than biological reaction.

The extent to which the chemical redox processes described herein can increase the solubility of trace metals (besides Fe and Mn) has not yet been established, but it is reasonable to expect that the dissolution of efficient metal scavengers such as Mn and Fe oxides will release significant levels of other metals into soil solution.

Fulvic and humic acids may play a role in reduction and dissolution of metals in addition to Fe and Mn in soils. Cases of Hg^{2+} reduction to Hg^0 (Alberts et al.,

1974) and Mo(VI) reduction to Mo(V) and Mo(III) (Goodman and Cheshire, 1982), have been reported to result from reaction with humic acids. There is evidence that the Hg reaction, at least, involves electron transfer between the metal and semiquinone-type radicals in the organic (Alberts et al., 1974). Strong complexation with functional groups in organic matter may stabilize certain oxidation states of metals, thereby inhibiting redox reactions that would normally proceed rapidly in homogeneous solution. A good example of this is the vanadyl cation, VO^{2+}, which is readily oxidized by O_2 in nonacidic solutions but persists for long periods when complexed to soil organic matter (McBride, 1978b). Redox processes at soil surfaces, then, are not controlled by the well-established equilibrium constants for reduction-oxidation in homogeneous solution, a fact that complicates prediction of the oxidation state and solubility of metals in soils.

All of the redox processes outlined, although often involving substrates of biological origin, are abiotic reactions. In the complex soil environment, it can become difficult to separate a biological process from one that is enabled by a biological process. However, sufficient numbers of examples of nonbiological electron-transfer processes have been reported to suggest that numerous redox reactions that involve metals in soils are abiotic. Although some of these reactions can only be activated in the presence of biological activity because of the requirement for oxidizable organic substrates, they need to be recognized as nonbiological processes with all of the characteristics of chemical reactions.

VI. Metal Adsorption by Organic Matter

Although metal bonding on organic matter can be viewed as an ion exchange process between H^+ and metal ions on acidic functional groups, the high degree of selectivity shown by organic matter for certain metals strongly suggests that some metals coordinate directly (i.e., form inner-sphere complexes) with the functional groups. A typical affinity sequence of organic matter for metals (at pH 5) is given in Table 3 (Schnitzer and Skinner, 1966, 1967), but these sequences

Table 3. Sequence of affinity of divalent metal ions for soil organic matter related to selected properties of the metals

Affinity sequence	Cu >	Ni >	Pb >	Co >	Ca >	Zn >	Mn >	Mg
Electronegativity (Pauling)	2.0	1.91	1.87	1.88	1.00	1.65	1.55	1.31
CFSE[a]	22.2	29.3	0	17.1	0	0	0	0
pK_1[b]	7.5	9.4	7.8	9.6	12.7	9.6	10.7	11.4

[a] Crystal field stabilization energy (kcal/mole) for an octahedral complex.
[b] Negative logarithm of the first hydrolysis constant.
Source: All numerical metal properties from Huheey (1972). Affinity sequence is from Stevenson and Ardakani (1972).

are commonly inconsistent, dependent on the nature of the organic matter, the method used to measure metal bonding, and the pH at which bonding is measured (Stevenson and Ardakani, 1972). Of particular concern is the fact that selectivity coefficients for metal adsorption are highly dependent on the degree of loading (i.e., quantity of metal adsorbed relative to quantity of adsorption sites) and the presence of competing metals (Hendrickson and Corey, 1981). At environmentally realistic metal loadings, small quantities of functional groups with specific preference for certain metals may heavily weight the selectivity coefficients in favor of adsorption of those metals. For example, at high adsorption levels, Cd^{2+} appears to have an affinity equal to Ca^{2+} for soil bonding sites. However, at low adsorption levels, Cd^{2+} is strongly preferred to Ca^{2+} (Hendrickson and Corey, 1981). It is known that Cd^{2+}, a relatively "soft acid" by the Pearson classification of Lewis acids, should prefer soft bases such as sulfur-containing ligands; thus, sulfhydryl groups in soil organics could generate very high preferences for trace levels of Cd^{2+} (and other soft acids such as Pb^{2+}). In contrast, a "hard acid" such as Ca^{2+} would preferentially bind ligands in the order O > N > S. Thus, Ca^{2+} could compete successfully with Cd^{2+} for the abundant carboxylate ligands in soils. Metals such as Cu^{2+} and Zn^{2+} are "borderline" acids, with intermediate behavior. For any given complexing ligand, the transition metals later in the first

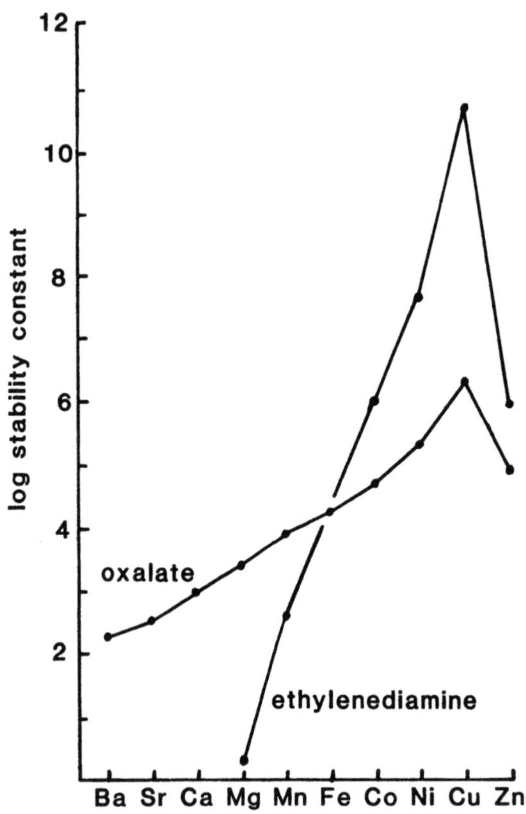

Figure 14. Stability of ethylenediamine and oxalate complexes with alkaline earth and transition metals, showing the Irving–Williams effect. (Adapted from Huheey, 1972).

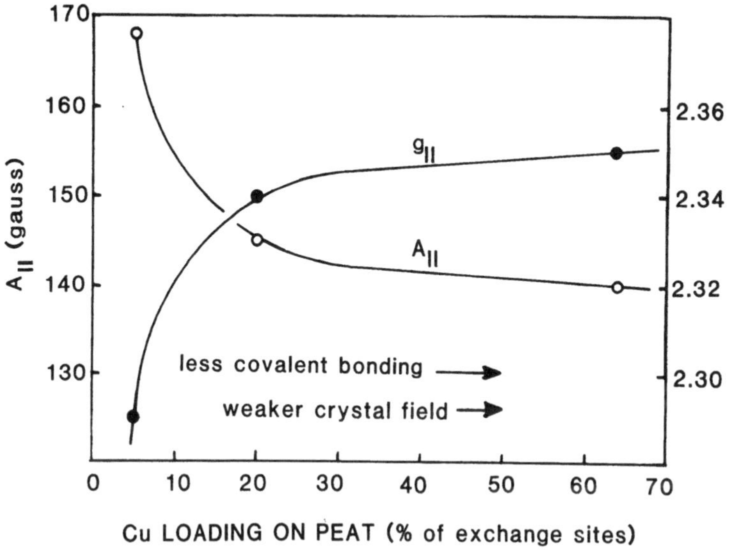

Figure 15. Hyperfine splitting (A_{11}) and g_{11} values for the ESR spectrum of Cu^{2+} bonded to a $Ca^{2\pm}$ peat at several loading levels. (Data from Baes, 1983).

row of the periodic table form increasingly stable complexes as a consequence of increasing metal acidity (decreasing radius) according to the Irving–Williams series. This is demonstrated in Figure 14 as an increasing relative stability of metal-amine complexes upon increasing the number of d-electrons and decreasing the ionic size of the metal. Superimposed on this ionic size effect is the "softness" factor, with the softer acids being those metals with greater numbers of d-electrons (e.g., Ni^{2+}, Cu^{2+}), and the harder acids occurring earlier in the transition metal series or in the alkaline earth group. Thus, Figure 14 reveals a strong relative preference of Cu^{2+} and Ni^{2+} for the softer N ligand (order of preference is S > N > O) and a preference of Mg^{2+}, Ca^{2+} and Mn^{2+} for the harder O ligand (order of preference is O > N > S).

There is evidence from ESR that low Cu^{2+} loading in organic soils favors more covalent bonding, which is revealed by a shift in *g*-values and hyperfine splitting as shown in Figure 15. This result has often been interpreted as complexation to amine-type N groups in preference to O-containing ligands. Low levels of Cu^{2+} complexed with organic matter produce g_{11} values below 2.30 (Lakatos et al., 1977b). In fact, small quantities of a Cu^{2+}-porphyrin complex ($g_{11} = 2.171$) have been detected in soil organic matter (Goodman and Cheshire, 1973). In contrast, Cu^{2+} bonded to organic matter at high loadings produces g_{11} values above 2.30, indicating that Cu^{2+} is largely coordinated to oxygen-containing ligands of a type that do not bond very covalently (McBride, 1978b; Boyd et al., 1981b). Although part of this Cu^{2+} is bonded rigidly, presumably as an inner-sphere complex, a higher degree of Cu^{2+} mobility is indicated at higher Cu^{2+} loadings. An illustration of the correlation between the rigid-limit g_{11} ESR parameter and the coordination environment of model and natural organics is seen in Figure 16. Here,

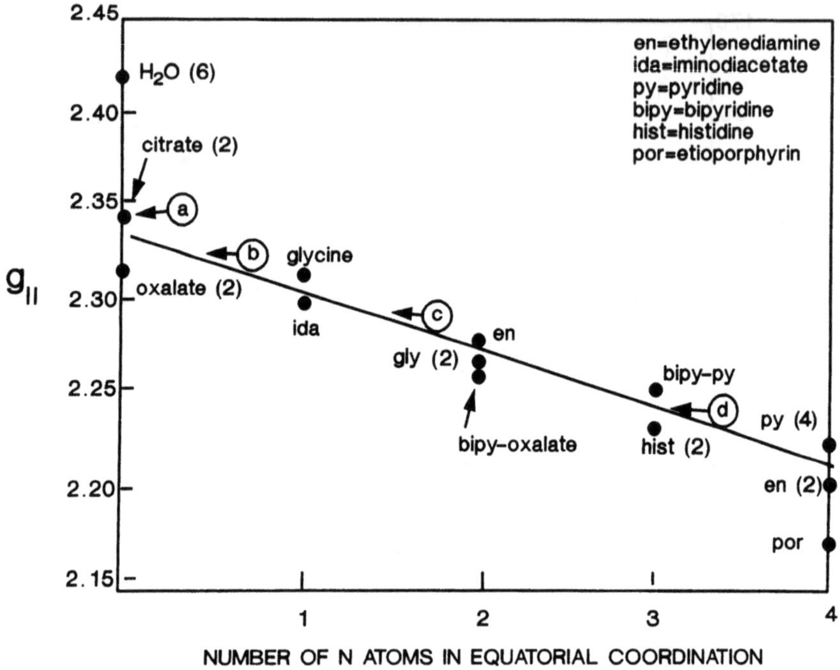

Figure 16. Correlation between g_{11} value of the rigid-limit ESR spectra and the number of N atoms in the equatorial plane of tetragonal Cu^{2+} complexes. The arrows indicate g_{11} values of: (a) a high-Cu humic acid (Boyd et al., 1981b). (b) a high-Cu humic acid (McBride, 1978b). (c) a low-Cu, low protein humic acid (Lakatos et al., 1977b). (d) a low-Cu, high protein humic acid (Lakatos et al., 1977b).
Parentheses denote number of molecules coordinated to Cu^{2+} (if this number is greater than one). (Additional spectral data from Kivelson and Neiman, 1961; McBride, 1985a).

high-Cu^{2+} organic matter (a and b) is observed to have a larger g_{11} value than low-Cu^{2+} organic matter (c and d), consistent with zero to one nitrogen ligands in the high-Cu^{2+} organics and one to three nitrogen ligands in the low-Cu^{2+} organics. ESR studies of Cu^{2+} bonding in a number of soil humic acids and melanins synthesized by soil fungi have produced a similar interpretation of bonding, usually involving one or two N ligands (Senesi et al., 1986, 1987). The difficulty with this interpretation, as will be seen later, is that some oxygen-containing ligands have the capability to bond covalently with Cu^{2+}, generating small values of g_{11} and large hyperfine splittings. Polyphenolic compounds, in particular, have this capability.

Potentiometric titration data for M^{2+} (M^{2+} = Ca^{2+}, Cd^{2+}, Zn^{2+}, Pb^{2+}, and Cu^{2+}) bonding in organic matter have been interpreted to indicate the formation of the following complex:

$$M^{2+} + A^{-} \rightleftharpoons MA^{+}$$

where A^{-} is the dissociated organic ligand (Marinsky et al., 1980). This interpretation contrasts with the common assumption that bonding of metals to organic matter is a chelation or complexation reaction with more than one functional group. Spectroscopic evidence reveals that *some* metals (e.g., Mn^{2+}) do *not* form inner-sphere complexes with organic matter (Alberts et al., 1976; Gamble et al., 1976; Deczky and Langford, 1978; McBride, 1978b). In fact, Cu^{2+}-humic acid complexes at fairly high-Cu^{2+} sorption levels are not particularly stable relative to Cu^{2+} chelates (see Figure 17) but comparable in stability to Cu-acetate. Infrared spectroscopy has indicated that Cu^{2+} at high levels in humic acid is bonded to carboxylate via a monodentate bond, but it does not determine whether a second functional group is involved in the bonding process (Boyd et al., 1981a). Inner-sphere complexation may give way to outer-sphere com-

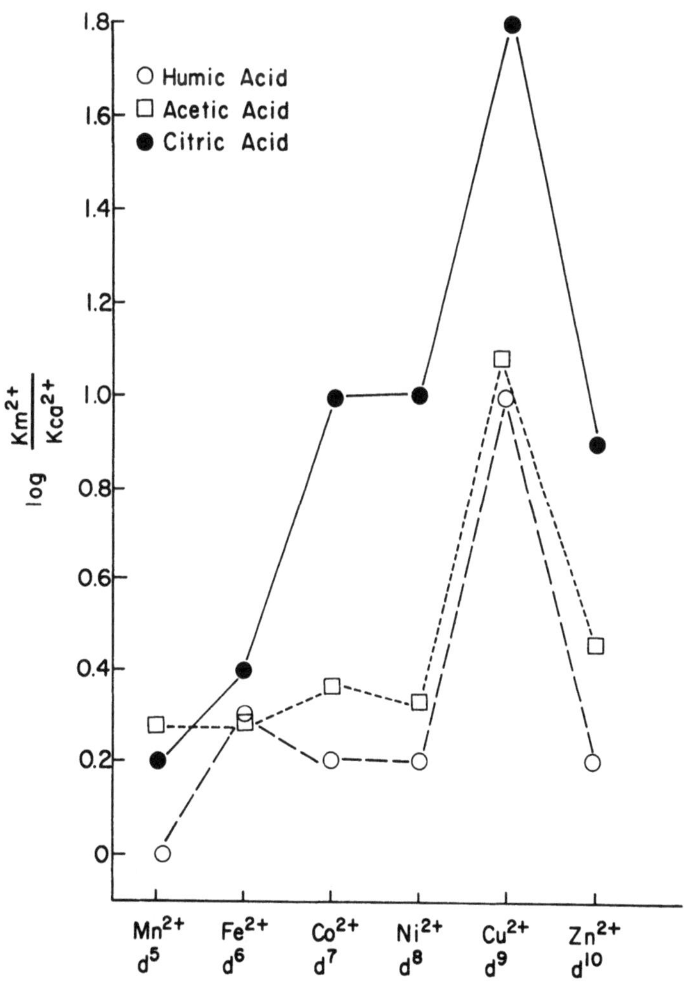

Figure 17. Relative bonding strength of transition metals (M^{2+}) for complexation to humic acid, acetic acid, and citric acid expressed as the ratio of the stability constants for the M^{2+} and Ca^{2+} complexes. (Bloom, 1978; reproduced with permission.)

plexation at high sorption level, with an equilibrium between inner- and outer-sphere coordination:

$$M(H_2O)_x^{2+} \cdots A^- \rightleftharpoons MA^- + xH_2O$$

which is sensitive to factors such as extent of site occupation by the metal, pH, and hydration state. The degree of site occupation affects the equilibrium because small quantities of a metal can bond at those sites with highest preference for the metal—these few sites may include highly selective chelating groups such as the porphyrin mentioned earlier. Higher pH generates a greater surface population of complexing ligands, A^-, and seems to favor the inner-sphere complexation of metals that are retained as hydrated ions at lower pH (Lakatos et al., 1977b; McBride, 1978b). Dehydration of metal-organic complexes tends to force the metal into direct coordination with organic ligands by removing the competing water. This conversion from largely outer-sphere complexation to inner-sphere complexation upon dehydration has been confirmed for $Fe^{2\pm}$ humic acid complexes using Mössbauer spectroscopy (Lakatos et al., 1977a). Thus, whether researchers observe chelates, inner-sphere complexes, or outer-sphere complexes depends to some degree on experimental conditions. Nevertheless, most evidence points to the tendency of Cu^{2+} and VO^{2+}, for example, to form inner-sphere complexes, and most of the other first-row transition metals (e.g., Mn^{2+}, Fe^{2+}, Co^{2+}) and alkaline earth metals (Ca^{2+}, Mg^{2+}) to form outer-sphere complexes. The failure of the selectivity coefficients for metal bonding to humic acid to follow the Irving–Williams series (Figure 17) supports this hypothesis. If all the metals represented in Figure 17 formed inner-sphere complexes, one could expect selectivity to increase with decreasing ionic radius (Bloom and McBride, 1979) as follows:

$$Mn^{2+} < Fe^{2+} < Co^{2+} < Ni^{2+} < Cu^{2+}$$

Transition-metal bonding on the most populous sites (carboxylic groups) of organic matter has many similarities to bonding on synthetic polycarboxylic compounds (McBride, 1980b). Cu^{2+} and VO^{2+} complexes of these compounds are clearly inner-sphere, whereas Mn^{2+} complexes tend to have intermediate behavior (Lakatos et al., 1977a, 1977b; Deiana et al., 1980; McBride, 1982b). In fact, the nonchelating polycarboxylic acids seem, on average, to represent metal-bonding behavior in soil organics more accurately than chelating molecules. One clear exception is the behavior of trivalent metals such as Fe^{3+}. The ESR signals of these Fe^{3+} complexes, although indicative of a low-symmetry (rhombic) ligand field (Senesi et al., 1986, 1987; McBride et al., 1983), do not specify the chemical nature of the ligands. The observed g-values near 4.2 are consistent with Fe^{3+} complexes with carboxylic acids as well as polyphenols (M.G. Johnson, 1986). Mössbauer spectra of $Fe^{3\pm}$ fulvic acid complexes are unable to further elucidate the nature of complexing ligands (Goodman and Cheshire, 1987). However, the observed pH stability of natural Fe^{3+}-organic complexes is best explained by a chelation mechanism that involves polyphenolic compounds (M.G. Johnson, 1986).

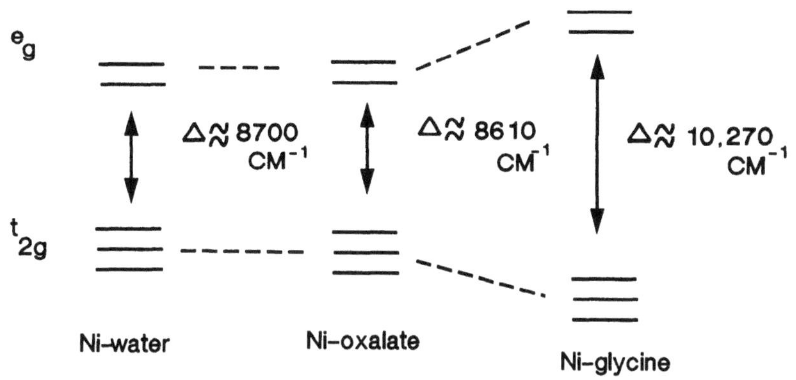

Figure 18. Energy levels of the 5 d-orbitals of Ni^{2+} in octahedral complexes with H_2O, oxalate, and glycine ligands. Estimates of the crystal-field splitting, Δ, are based on data from Huheey, 1972.

In general, the more electronegative metal ions bond most strongly to organic matter (Table 3), evidence that the metal-ligand bond has significant covalency. Although the crystal field strength of metals bonded to carboxylic groups of organic matter is probably slightly greater than the crystal-field strength of the hydrated metal, the driving force for transition-metal bonding by organic matter is unlikely to be largely crystal-field stabilization energy. Specific evidence for this is seen with the Ni^{2+} ion, which would gain the most stability from this (Table 3) and is complexed more strongly to amines than to carboxylic groups (Figure 14), yet Cu^{2+} is more strongly bonded than Ni^{2+} on organic matter and polycarboxylic acids. Figure 18 reveals essentially no stabilization of Ni^{2+} upon complexation of the aqueous ion with carboxylate anions of oxalate, but a very significant stabilization upon complexation with glycine. For Ni^{2+}, the crystal-field stabilization energy in weak octahedral ligand fields is given by $-1.2\Delta+3P$ where Δ and P are crystal-field splitting and electron-pairing energy, respectively. Since P has the same value for all three complexes depicted in Figure 18, the oxalate complex is calculated to be destabilized by $(-1.2)(8610-8700)\text{cm}^{-1} = +108\ \text{cm}^{-1} = 0.31$ kcal/mole, a negligible factor in complexation. In contrast, the glycine complex is stabilized by 5.4 kcal/mole relative to the hydrated metal ion. The driving "force" for metal adsorption, then, can be the higher covalency of the metal-organic bond relative to the metal-H_2O bond (as in the case of glycine) and/or the opportunity for multidentate coordination of the metal (as in the case of oxalate). Multidentate bonding, including chelation, provides additional stability for metal-organic complexes by increasing the total entropy of the system. The qualitatively good correlation between electronegativity and bonding preference, revealed in Table 3, is evidence for the importance of covalent bonding. Although the ESR spectra of Cu^{2+} bonded in organic matter suggest the involvement of N-ligands, this is evidently insufficient to cause Ni^{2+} to be complexed preferentially to Cu^{2+}. It appears likely that the high loadings of metals commonly used to establish metal-affinity sequences saturate these ligands, causing the carboxylate groups to determine the apparent order of metal preference.

Desorption of transition and heavy metals from organic matter, a process involving displacement of organic ligands by water, has not often been studied. Data on rates of adsorption and desorption in peat indicate that those metals that bond strongly in organic matter (e.g., Pb^{2+}, Cu^{2+}) are most rapidly adsorbed and most slowly desorbed (Bunzl et al., 1976). Because greater heats of adsorption are expected for the inner-sphere complexes formed between organic ligands and metals such as Cu^{2+}, desorption necessarily requires that a large activation energy be overcome. Isotope exchange data reveal that the bulk of the adsorbed Cu^{2+} and Fe^{3+} on organic matter is not labile over a period of one day (McLaren and Crawford, 1974; Sedlacek et al., 1987). Rates of formation of metal-ligand complexes are known to be strongly correlated to the rates at which inner-sphere water molecules are exchanged on metal hydrates (Cotton and Wilkinson, 1980). $Cr(H_2O)_6{}^{3+}$ is an example of an aquocomplex with very slow H_2O exchange rates and extremely sluggish exchange with ligands. The unusually large crystal-field stabilization energy of Cr^{3+} presents an energy barrier to ligand exchange because the exchange process necessitates that the six-coordinate complex be distorted. $Al(H_2O)_6{}^{3+}$ is somewhat more labile, but ligand exchange reactions involving Al^{3+}-organic complexes in soil solution are measurably slow, on the order of minutes to hours (B.R. James et al., 1983). In general, metals within the same group in the periodic table and having the same charge tend to have more rapid rates of exchange as their ionic radii increase. However, some metals do not follow the radius rule very well. For example, Cu^{2+} has an unusually rapid exchange rate for ligands relative to the other first-row transition metals (Cotton and Wilkinson, 1980). The Jahn–Teller distortion of octahedral Cu^{2+} complexes may account for this lability because the axial ligands are bonded much less strongly than the equatorial ligands. The presence of ligands other than H_2O on hydrated metals can catalyze ligand exchange, as is demonstrated by the fact that $Fe(H_2O)_5OH^{2+}$ forms complexes with organic ligands much more rapidly than $Fe(H_2O)_6{}^{3+}$ (Stone, 1986). This may be an effect of charge reduction at the metal by more covalent bonding in Fe-hydroxy than Fe-H_2O complexes.

Future research may better define the role of lability relative to thermodynamic stability of metal-organic complexes in understanding metal behavior in soil organic matter. At the present time, however, both factors must be taken into account in studies of metal adsorption by soil organic matter within a limited time frame.

VII. Speciation of Metals in Solution

All of the sorption/desorption and precipitation/dissolution mechanisms described herein are sensitive to the chemical form of the dissolved metal. Generally, soluble organics retain metals in solution in nonadsorbing forms (usually as anions or neutral species) and can therefore raise the total dissolved

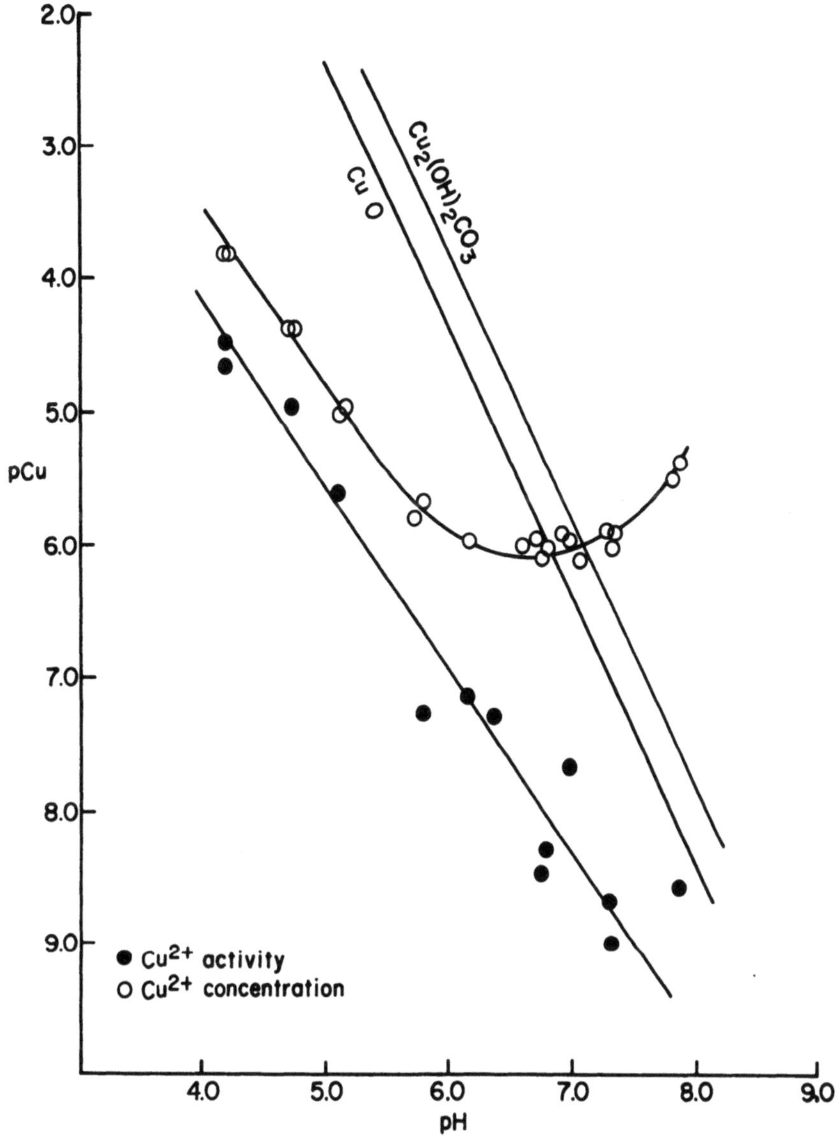

Figure 19. Activity and concentration of Cu^{2+} in soil solution as a function of pH after adding 40 ppm Cu^{2+} to a mineral soil. Solubility lines for Cu precipitates assume atmospheric level of CO_2.

metal concentration above that expected if only the free hydrated metal were present. Such a result is demonstrated for Cu^{2+} in Figure 19, where the total dissolved Cu is greater than free $Cu(H_2O)_6^{2+}$ in soil solution, especially at higher pH. Therefore, although raising soil pH reduces the concentration of *free* metal by adsorption and precipitation phenomena, there may be much less effect on total dissolved metal because the higher pH promotes dissolution of soil organics and formation of soluble metal-organic complexes. Some metals, such as Fe^{3+}, have extremely low solubility as the free metal at common soil pH values, yet significant concentrations of total dissolved Fe are detected in many soil solutions. For example, cathodic stripping voltammetry has been used to detect levels of total Fe in acid soil solutions in the range of 2 to 20 μM, but only about 1% of this was found to be in a labile (free?) form (B.R. James and Bouldin, 1986). The ESR investigations of soil solutions and of model Fe^{3+}-organic complexes suggest that Fe is likely to be maintained in a soluble form by o-diphenolic or polyphenolic compounds that chelate Fe^{3+} (M.G. Johnson, 1986). Most other chelating agents investigated failed to demonstrate stability comparable to the natural Fe^{3+} complex. Based on existing evidence, then, it appears that dissolved metals such as Fe^{3+} and Cu^{2+} are almost completely in organically complexed form (Hodgson et al., 1965; McBride and Blasiak, 1979; B.R. James and Bouldin, 1986). However, as can be deduced from Figure 19, the fraction of Cu^{2+} in the complexed relative to free form diminishes at lower pH, and this type of behavior is likely for most complexing metals. Metals such as Co^{2+}, Cd^{2+}, and Zn^{2+}, because of their lesser tendencies to form soluble complexes with oxygen-containing functional groups of soluble organic matter, are complexed to a smaller degree in soil solution (Hodgson et al., 1965; McBride and Blasiak, 1979; Tyler and McBride, 1982). For example, between 13% and 43% of Zn in acid soil solutions has been reported in exist in labile form (B.R. James and Bouldin, 1986). Most of the labile Zn is believed to exist as the uncomplexed Zn^{2+} cation, but there is experimental uncertainty associated with this and all other presently available methods of metal speciation. Anomalies observed in Langmuir adsorption isotherms for Zn^{2+} added to soils (Pulford, 1986) are consistent with the hypothesis that a small fraction of Zn^{2+} added to soils enters into a soluble *nonadsorbing* form, most likely a complex with fulvic acid. This nonadsorbing fraction produces a discontinuity in the linear form of the Langmuir equation at low Zn concentrations.

The importance of soluble organically complexed forms of metals in controlling availability for plant uptake cannot be overemphasized. Recent studies have revealed that although estimates of free Cu^{2+} in soil solution can be correlated to absorption of Cu^{2+} by roots, additions of large quantities of organic matter to the soil alter this correlation, which results in greater uptake by roots at a given concentration of free Cu^{2+} (Minnich et al., 1987). These results support the concept of soluble organics as "metal carriers."

The types of organic molecules involved in forming soluble complexes are likely to include oxalic, citric, malic, tartaric, and many other organic acids, aliphatic and aromatic, that are common in soil solution (Stevenson and Ardakani, 1972). The relative importance of these different organic acids depends on the complexing metal and the pH, as is illustrated in Figure 20. In this hypothetical example, 10^{-5} *M* total Cu and Zn are simultaneously present in solution with 10^{-4} *M* catechol and salicylic acid. Speciation calculations predict that an extremely small fraction of total soluble Zn actually forms complexes with these ligands at low pH, with salicylate dominating the speciation. For Cu^{2+}, important differences are noted; the fraction of the metal complexed by organics

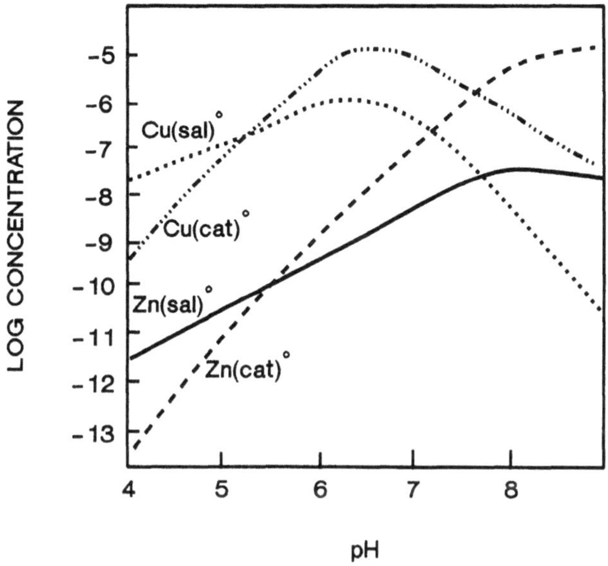

Figure 20. Predicted concentrations of metal-ligand complexes as a function of pH in a solution containing 10^{-5} *M* total Cu and Zn as well as 10^{-4} *M* catechol and salicylate. Metal-hydroxy and metal-multiligand complexes are not plotted to keep the diagram simple. Data are calculated using the TITRATOR microcomputer program (Cabaniss, 1987). No precipitation of metal hydroxides is predicted to occur from the calculated activities of the free metals.

is much higher at low pH, and catechol is a weaker ligand than salicylate for pH values below 5.5. For both Cu^{2+} and Zn^{2+}, the speciation shifts in favor of catechol complexation at higher pH, and the fraction of Zn complexed with organics becomes significant above pH 7.0. As Figure 20 does not depict the metal-ligand species involving two organic ligands per metal (for sake of simplicity), it appears to show a decrease in the level of Cu complexed by catechol at high pH. This is a result of the dominance of the $Cu(catechol)_2^{2-}$ species over the $Cu(catechol)^0$ species above pH 7.6, and the Cu is, in fact, predicted to be almost totally complexed with catechol at high pH.

The ESR spectra of Cu^{2+}-salicylate and Cu^{2+}-catechol solutions confirm the existence of two complexes believed to be the CuL^0 and CuL_2^{2-} species (where L = deprotonated ligand). In the case of salicylate, the spectrum of one species is evident in the pH range 4.0 to 7.0, whereas the other becomes apparent above pH 7.0. With catechol, the spectrum of the first complex with Cu^{2+} is not evident until the pH exceeds 4.0, whereas the second complex forms below pH 7.0 and appears stable to pH 9.0 or higher. The ESR parameters of these complexes, plotted in Figure 21, indicate that increased organic-ligand coordination to the metal reduces the *g*-value and increases the hyperfine splitting (*A*-value) relative to $Cu(H_2O)_6^{2+}$, a trend that can be interpreted to indicate increased covalent bonding in the metal-ligand complex (Kivelson and Neiman,

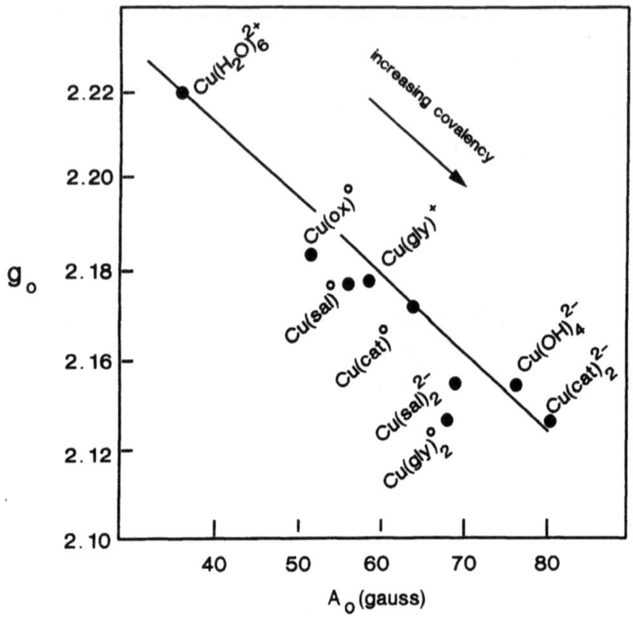

Figure 21. Correlation between *g*-values and hyperfine splittings (*A*-values) of selected soluble Cu-ligand complexes.

1961). The somewhat larger effect of catecholate versus salicylate in this regard may indicate that the former ligand is a better electron donor, and the greater stability constant for the catecholate complex suggests this. A comparison of the structure of these chelates

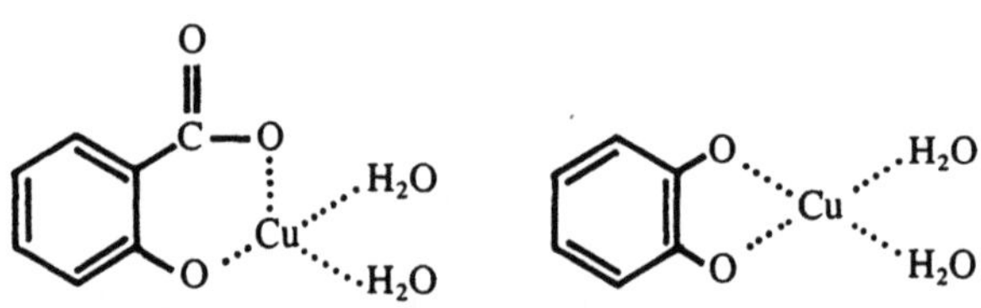

(A) (B)

reveals that the π-electron density of the aromatic ring can be more effectively donated to the electronegative metal ion in structure B.

Simple complexes of the salicylate or o-diphenol type have not been directly identified in soil solutions, although salicylate is commonly mentioned as a likely ligand "type" in soluble and insoluble organic matter of soil. The rigid-limit ESR spectra of Cu^{2+} bound to soil organic matter at relatively high levels produces *g*-values that, when averaged by the equation $g_0 = \frac{1}{3}g_{11} + \frac{2}{3}g_{\perp}$, are in the range for Cu(salicylate)° or Cu(catechol)° complexes (McBride, 1978b; Bloom and McBride, 1979; Boyd et al., 1981b). On the other hand, Cu^{2+} bonded to organic matter at low levels produces lower *g*-values and larger hyperfine splittings (Lakatos et al., 1977b), which Figure 21 indicates would not be inconsistent with

Cu^{2+} coordination to two bidentate O-containing ligands. However, as discussed earlier, these low *g* and high *A* values can also be caused by the increased covalent bonding associated with N-ligand coordination to Cu^{2+}. Even this general trend is not reliable; for example, Figure 21 reveals that Cu-glycine complexes do not have substantially different ESR parameters from Cu-salicylate complexes despite the coordination of N to Cu^{2+} in the former case. Thus, although the ESR parameters reflect the degree of covalency of the metal-ligand bond (Kivelson and Neiman, 1961), they do not reliably identify the ligand types involved.

It does not appear at this stage that the identity of organic functional groups involved in complexing metals in the solution (or solid) phase of soils can be stated with any certainty. Although the pattern of complexation shown in Figure 20 for model complexes is qualitatively similar to that seen in soil solutions, with Cu complexed to a greater degree than Zn at low pH and both metals complexed to a high degree at high pH, the quantitative estimates of metal complexation predicted by these model compounds is lower than actually observed. Speciation calculations (not shown) indicate that strongly acidic organic acids that are good chelating ligands (e.g., oxalate) could explain the higher levels of metal complexation observed in soil solution, as could N-containing ligands such as amino acids. Until the particular ligand types involved in metal complexation in soil solution are identified to some degree, speciation models that purport to predict the degree of metal complexation in undefined soil solutions must be treated with some caution.

VIII. Summary

Given the complexity of metal interactions with other metals and with soil surfaces described in this chapter, it is evident that accounting for all processes that could determine the solubility of trace metals in soils is beyond the capability of any single theoretical approach or model. Although this fact would appear to force a retreat to the empirical methods for description and prediction of metal solubility, it is clear that a theoretical framework must be maintained and built upon if any advances in understanding are to be achieved.

References

Adamson, A.W. 1976. *Physical chemistry of surfaces*, 3d ed. Wiley, NY.

Alberts, J.J., J.E. Schindler, R.W. Miller, and D.E. Nutter. 1974. Elemental mercury evolution mediated by humic acid. *Science* 184:895–897.

Alberts, J.J., J.E. Schindler, D.E. Nutter, and E. Davis. 1976. Elemental infrared spectrophotometric and electron spin resonance investigations of non-chemically isolated humic material. *Geochim. Cosmochim. Acta* 40:369–372.

Baes, A.U. 1983. Exchange of divalent cations in an acid-washed peat and a polycarboxylate resin. M.S. thesis, Univ. of Minnesota.

Bartlett, R., and B. James. 1979. Behavior of chromium in soils. III. Oxidation. *J. Environ. Qual.* 8:31–35.

Benjamin, M.M. 1983. Adsorption and surface precipitation of metals on amorphous iron oxyhydroxide. *Environ. Sci. Technol.* 17:686–692.

Benjamin, M.M., and J.O. Leckie. 1982. Effects of complexation by Cl, SO_4, and S_2O_3 on adsorption behavior of Cd on oxide surfaces. *Environ. Sci. Technol.* 16:162–170.

Bleam, W.F., and M.B. McBride. 1985. Cluster formation versus isolated-site adsorption. A study of Mn(II) and Mg(II) adsorption on boehmite and goethite. *J. Colloid Interfac. Sci.* 103:124–132.

Bleam, W.F., and M.B. McBride. 1986. The chemistry of adsorbed Cu(II) and Mn(II) in aqueous titanium dioxide suspensions. *J. Colloid Interfac. Sci.* 110:335–346.

Bloom, P.R. 1978. Exchange of hydrogen, aluminum and other metal ions in soil organic matter and acid soils. Ph.D. diss., Cornell Univ.

Bloom, P.R., and M.B. McBride. 1979. Metal ion binding and exchange with hydrogen ions in acid-washed peat. *Soil Sci. Soc. Am. J.* 43:687–692.

Bloom, P.R., M.B. McBride, and B. Chadbourne. 1977. Adsorption of aluminum by a smectite: I. Surface hydrolysis during Ca^{2+}-Al^{3+} exchange. *Soil Sci. Soc. Am. J.* 41:1068–1073.

Bolland, M.D.A., A.M. Posner, and J.P. Quirk. 1977. Zinc adsorption by goethite in the absence and presence of phosphate. *Aust. J. Soil Res.* 15:279–286.

Bourg, A.C.M., S. Joss, and P.W. Schindler. 1979. Ternary surface complexes. 2. Complex formation in the system silica-Cu(II)-2,2'-*bipyridyl. Chimia* 33:19–21.

Bourg, A.C.M., and P.W. Schindler. 1979. Effect of ethylenediaminetetraacetic acid on the adsorption of copper(II) at amorphous silica. *Inorg. Nucl. Chem. Lett.* 15:225–229.

Bowden, J.W., A.M. Posner, and J.P. Quirk. 1977. Ionic adsorption on variable charge mineral surfaces. Theoretical-charge development and titration curves. *Aust. J. Soil Res.* 15:121–136.

Boyd, S.A., L.E. Sommers, and D.W. Nelson. 1981a. Copper(II) and iron(III) complexation by the carboxylate group of humic acid. *Soil Sci. Soc Am. J.* 45:1241–1242.

Boyd, S.A., L.E. Sommers, D.W. Nelson, and D.K. West. 1981b. The mechanism of copper(II) binding by humic acid: An electron spin resonance study of a copper(II)-humic acid complex and some adducts with nitrogen donors. *Soil Sci. Soc. Am. J.* 45:745–749.

Bricker, O. 1965. Some stability relations in the system Mn-O_2-H_2O at 25° and one atmosphere total pressure. *Am. Mineral.* 50:1296–1354.

Brümmer, G., K.G. Tiller, U. Herms, and P.M. Clayton. 1983. Adsorption-desorption and/or precipitation–dissolution processes of zinc in soils. *Geoderma* 31:337–354.

Bunzl, K., W. Schmidt, and B. Sansoni. 1976. Kinetics of ion exchange in soil organic matter. IV. Adsorption and desorption of Pb^{2+}, Cu^{2+}, Cd^{2+}, Zn^{2+}, and Ca^{2+} by peat. *J. Soil Sci.* 27:32–41.

Cabaniss, S.E. 1987. Titrator: An interactive program for aquatic equilibrium calculations. *Environ. Sci. Technol.* 21:209–210.

Carrington, A., and A.D. McLachlan. 1967. *Introduction to magnetic resonance*. Harper & Row, NY.

Cavallaro, N., and M.B. McBride. 1978. Copper and cadmium adsorption characteristics of selected acid and calcareous soils. *Soil Sci. Soc. Am. J.* 42:550–556.

Chesworth, W., and J. Dejou. 1980. Are considerations of mineralogical equilibrium relevant to pedology? Evidence from a weathered granite in central France. *Soil Sci.* 130:290–292.

Clark, C.J., and M.B. McBride. 1984. Chemisorption of Cu(II) and Co(II) on allophane and imogolite. *Clays Clay Miner.* 32:300–310.

Clark, C.J., and M.B. McBride. 1985. Adsorption of Cu(II) by allophane as affected by phosphate. *Soil Sci.* 139:412–421.

Clementz, D.M., T.J. Pinnavaia, and M.M. Mortland. 1973. Stereochemistry of hydrated copper(II) ions on the interlamellar surfaces of layer silicates. An electron spin resonance study. *J. Phys. Chem.* 77:196–200.
Cornell, R.M., and R. Giovanoli. 1987. Effect of manganese on the transformation of ferrihydrite into goethite and jacobsite in alkaline media. *Clays Clay Miner.* 35: 11–20.
Cotton, F.A., and G. Wilkinson. 1980. *Advanced inorganic chemistry*, 4th ed. Wiley, NY.
Davies, S.H.R. 1986. Mn(II) oxidation in the presence of lepidocrocite: The influence of other ions. pp. 487–502 In: *Geochemical processes at mineral surfaces*, J.A. Davis and K.F. Hayes (eds.). ACS Symposium Series no. 323, ACS, Washington, DC.
Davis, J.A., C.C. Fuller, and A.D. Cook. 1987. A model for trace metal sorption processes at the calcite surface: Adsorption of Cd^{2+} and subsequent solid solution formation. *Geochim. Cosmochim. Acta* 51:1477–1490.
Davis, J.A., and J.O. Leckie. 1978. Effect of adsorbed complexing ligands on trace metal uptake by hydrous oxides. *Environ. Sci. Technol.* 12:1309–1315.
Deczky, K., and C.H. Langford. 1978. Application of water nuclear magnetic relaxation times to study of metal complexes of the soluble soil organic fraction fulvic acid. *Can. J. Chem.* 56:1947–1951.
Deiana, S., L. Erre, G. Micera, P. Piu, and C. Gessa. 1980. Coordination of transition-metal ions by polygalacturonic acid: A spectroscopic study. *Inorg. Chim. Acta* 46:249–253.
Dillard, J.G., and C.V. Schenck. 1986. Interaction of Co(II) and Co(III) complexes on synthetic birnessite: surface characterization. pp. 503–522. In: *Geochemical processes at mineral surfaces*, J.A. Davis and K.F. Hayes (eds.). ACS Symposium Series no. 323, ACS, Washington, DC.
DiToro, D.M., J.D. Mahony, P.R. Kirchgraber, A.L. O'Byrne, L.R. Pasquale, and D.C. Piccirilli. 1986. Effects of nonreversibility, particle concentration and ionic strength on heavy metal sorption. *Environ. Sci. Technol.* 20:55–61.
Driessens, F.C.M. 1986. Ionic solid solutions in contact with aqueous solutions. pp. 524–560. In: *Geochemical processes at mineral surfaces*, J.A. Davis and K.F. Hayes (eds.). ACS Symposium Series no. 323, ACS, Washington, DC.
Dugger, D.L., J.H. Stanton, B.N. Irby, B.L. McConnell, W.W. Cummings, and R.W. Maatman. 1964. The exchange of twenty metal ions with the weakly acidic silanol group of silica gel. *J. Phys. Chem.* 68:757–760.
Ebinger, M.H., and D.G. Schulze. 1986. Properties of Mn-substituted magnetic iron oxides. *Agron. Abstr.*, Am. Soc. Agron., Madison, WI, p. 274.
Elliot, H.A., and C.P. Huang. 1979. The effect of complex formation on the adsorption characteristics of heavy metals. *Environ. Int.* 2:145–155.
Elliot, H.A., and C.P. Huang. 1981. Adsorption characteristics of some Cu(II) complexes on aluminosilicates. *Water Res.* 15:849–855.
el-Sayed, M.H., R.G. Burau, and K.L. Babcock. 1970. Thermodynamics of copper(II)-calcium exchange on bentonite clay. *Soil Sci. Soc. Am. Proc.* 34:397–400.
Farley, K.J., D.A. Dzombak, and F.M.M. Morel. 1985. A surface precipitation model for the sorption of cations on metal oxide. *J. Colloid Interfac. Sci.* 106:226–242.
Farrah, H., and W.F. Pickering. 1976. The sorption of copper species by clays. II. Illite and montmorillonite. *Aust. J. Chem.* 29:1177–1184.
Farrah, H., and W.F. Pickering. 1977. The sorption of lead and cadmium species by clay minerals. *Aust. J. Chem.* 30:1417–1422.
Fernandez, I.J., and P.A. Kosian. 1987. Soil air carbon dioxide concentrations in a New England spruce-fir forest. *Soil Sci. Soc. Am. J.* 51:261–263.

Forbes, E.A., A.M. Posner, and J.P. Quirk. 1976. The specific adsorption of divalent Cd, Co, Cu, Pb, and Zn on goethite. *J. Soil Sci.* 27:154–166.

Gadde, R.R., and H.A. Laitinen. 1974. Studies of heavy metal adsorption by hydrous iron and manganese oxides. *Anal. Chem.* 46:2022–2026.

Gamble, D.S., C.P. Langford, and J.P.K. Tong. 1976. The structure and equilibria of a manganese(II) complex of fulvic acid studied by ion exchange and nuclear magnetic resonance. *Can. J. Chem.* 54:1239–1245.

Garcia-Miragaya, J., and A.L. Page. 1976. Influence of ionic strength and inorganic complex formation on the sorption of trace amounts of Cd by montmorillonite. *Soil Sci. Soc. Am. J.* 40:658–663.

Garcia-Miragaya, J., and A.L. Page. 1977. Influence of exchangeable cation on the sorption of trace amounts of cadmium by montmorillonite. *Soil Sci. Soc. Am. J.* 41:718–721.

Gerth, J., and G. Brümmer. 1981. Effect of temperature and reaction time on the adsorption of nickel, zinc, and cadmium by goethite. *Mitteilgn. Dtsch. Bodenkundl. Gesellsch.* 32:229–238.

Gerth, J., and G. Brümmer. 1983. Adsorption und Festlegung von Nickel, Zink und Cadmium durch Goethit (α-FeOOH). *Fresenius Z. Anal. Chem.* 316:616–620.

Gilmour, J.T., and J.A. Kittrick. 1979. Solubility and equilibria of zinc in a flooded soil. *Soil Sci. Soc. Am. J.* 43:890–892.

Goldberg, S. 1985. Chemical modelling of anion competition on goethite using the constant capacitance model. *Soil Sci. Soc. Am. J.* 49:851–856.

Golden, D.C., J.B. Dixon, and C.C. Chen. 1986. Ion exchange, thermal transformations, and oxidizing properties of birnessite. *Clays Clay Miner.* 34:511–520.

Goodman, B.A., and M.V. Cheshire. 1973. Electron paramagnetic resonance evidence that copper is complexed in humic acid by the nitrogen of porphyrin groups. *Nature New Biol.* 244:58–159.

Goodman, B.A., and M.V. Cheshire. 1982. Reduction of molybdate by soil organic matter: EPR evidence for formation of both Mo(V) and Mo(III). *Nature* 299:618–620.

Goodman, B.A., and M.V. Cheshire. 1987. Characterization of iron-fulvic acid complexes using Mossbauer and EPR spectroscopy. *Sci. Total Environ.* 62:229–240.

Hathaway, B.J., and C.E. Lewis. 1969a. Electronic properties of transition-metal complex ions adsorbed on silica gel. Part I. Nickel(II) complexes. *J. Chem. Soc., A* 1176–1183.

Hathaway, B.J., and C.E. Lewis. 1969b. Electronic properties of transition-metal complex ions adsorbed on silica gel. Part II. Cobalt(II) and Cobalt(III). *J. Chem. Soc., A* 1183–1188.

Hayes, K.F., and J.O. Leckie. 1986. Mechanism of lead ion adsorption at the goethite-water interface. pp. 114–141. In: *Geochemical processes at mineral surfaces*, J.A. Davis and K.F. Hayes (eds.). ACS Symposium Series no. 323, ACS, Washington, DC.

Healy, T.W., R.O. James, and R. Cooper. 1968. The adsorption of aqueous Co(II) at the silica-water interface. pp. 62–73. In: *Adsorption from aqueous solution*, W.J. Weber and E. Matijevic (eds.). Advan. in Chem. Ser. no. 79, ACS, Washington, DC.

Helyar, K.R., D.N. Munns, and R.G. Burau. 1976. Adsorption of phosphate by gibbsite. II. Formation of a surface complex involving divalent cations. *J. Soil Sci.* 27:315–323.

Hem, J.D. 1978. Redox processes at surfaces of manganese oxide and their effects on aqueous metal ions. *Chem. Geol.* 21:199–218.

Hendrickson, L.L., and R.B. Corey. 1981. Effect of equilibrium metal concentrations on apparent selectivity coefficients of soil complexes. *Soil Sci.* 131:163–171.

Herms, U., and G. Brümmer. 1984. Einflussgrossen der Schwermetattöslichkeit und-bindung in Böden. *Z. Pflanzenernaehr. Bodenk.* 147:400–424.

Hider, R.C., A.R. Mohd-Nor, J. Silver, I.E.G. Morrison, and L.V.C. Rees. 1981. Model compounds for microbial iron-transport compounds. Part 1. Solution chemistry and Mössbauer study of iron(II) and iron(III) complexes from phenolic and catecholic systems. *J. Chem. Soc., Dalton* 609–622.

Hodges, S.C., and L.W. Zelazny. 1983. Interactions of dilute, hydrolyzed aluminum solutions with clays, peat, and resin. *Soil Sci. Soc. Am. J.* 47:206–212.

Hodgson, J.F. 1960. Cobalt reactions with montmorillonite. *Soil Sci. Soc. Am. Proc.* 24:165–168.

Hodgson, J.F., H.R. Geering, and W.A. Norvell. 1965. Micronutrient cation complexes in soil solution: Partition between complexed and uncomplexed forms by solvent extraction. *Soil Sci. Soc. Am. Proc.* 29:665–669.

Hodgson, J.F., K.G. Tiller, and M. Fellows. 1964. The role of hydrolysis in the reaction of heavy metals with soil-forming materials. *Soil Sci. Soc. Am. Proc.* 28:42–46.

Hsu, P.H. 1977. Aluminum hydroxides and oxyhydroxides. pp. 99–143. *In* "Minerals in soil environments," J.B. Dixon and S.B. Weed (eds.). Soil Sci. Soc. Am., Madison, WI.

Huheey, J.E. 1972. *Inorganic chemistry: Principles of structure and reactivity.* Harper & Row, NY.

Hutcheon, A.T. 1966. Thermodynamics of cation exchange on clay: Ca-K-montmorillonite. *J. Soil Sci.* 17:339–355.

Inskeep, W.P., and J. Baham. 1983. Adsorption of Cd(II) and Cu(II) by Na-montmorillonite at low surface coverage. *Soil Sci. Soc. Am. J.* 47:660–665.

Iu, K.L., I.D. Pulford, and H.J. Duncan. 1981a. Influence of waterlogging and lime or organic matter additions on the distribution of trace metals in an acid soil. I. Manganese and iron. *Plant and Soil* 59:317–326.

Iu, K.L., I.D. Pulford, and H.J. Duncan. 1981b. Influence of waterlogging and lime or organic matter additions on the distribution of trace metals in an acid soil. II. Zinc and copper. *Plant and Soil* 59:327–333.

Jahiruddin, M., B.J. Chambers, N.T. Livesey, and M.S. Cresser. 1986. Effect of liming on extractable Zn, Cu, Fe, and Mn in selected Scottish soils. *J. Soil Sci.* 37:603–615.

James, B.R. and D.R. Bouldin. 1986. A cathodic stripping voltammetric method for nanomolar concentrations of labile and total iron and zinc in soil solutions. *Comm. Soil Sci. Plant Anal.* 17(11):1185–1201.

James, B.R., C.J. Clark, and S.J. Riha. 1983. An 8-hydroxyquinoline method for labile and total aluminum in soil extracts. *Soil Sci. Soc. Am. J.* 47:893–897.

James B.R., and S.J. Riha. 1984. Soluble aluminum in acidified organic horizons of forest soils. *Can. J. Soil Sci.* 64:637–646.

James B.R., and S.J. Riha. 1986. pH buffering in forest soil organic horizons: Relevance to acid precipitation. *J. Environ. Qual.* 15:229–234.

James, R.O., and T. W. Healy. 1972a. Adsorption of hydrolyzable metal ions at the oxide-water interface. I. Co(II) adsorption on SiO_2 and TiO_2 as model systems. *J. Colloid Interfac. Sci.* 40:42–52.

James, R.O., and T.W. Healy. 1972b. Adsorption of hydrolyzable metal ions at the oxide-water interface. III. A thermodynamic model of adsorption. *J. Colloid Interfac. Sci.* 40:65–81.

Jauregui, M.A., and H.M. Reisenauer. 1982. Dissolution of oxides of manganese and iron by root exudate components. *Soil Sci. Soc. Am. J.* 46:314–317.

Jenne, E.A. 1968. Controls on Mn, Fe, Co, Ni, Cu, and Zn concentrations in soils and water: the significant role of hydrous Mn and Fe oxides. pp. 337–387. *In Trace inorganics in water.* Advanc. in Chem. Ser. no. 73. ACS, Washington, DC.

Johnson, L.J. and C.H. Chu. 1983. Mineralogical characterization of selected soils from northeastern United States. Bulletin 847, Ag. Expt. Sta., Pennsylvania State Univ.

Johnson, M.G. 1986. Clay mineralogy and chemistry of selected Adirondack spodosols. Ph.D. diss., Cornell Univ.

Jorgensen, S.S. 1976. Dissolution kinetics of silicate minerals in aqueous catechol solutions. *J. Soil Sci.* 27:183–195.

Jurinak, J.J., and N. Bauer. 1956. Thermodynamics of zinc adsorption on calcite, dolomite and magnesite-type minerals. *Soil Sci. Soc. Am. Proc.* 20:466–471.

Keizer, P., and M.G.M. Bruggenwert. 1986. Sorption of heavy metals by clay-aluminum hydroxide-complexes. Proc. NATO symposium on Soil Colloid-Soil Solution Interface, Ghent, Belgium.

Kinniburgh, D.G., M.L. Jackson, and J.K. Syers. 1976. Adsorption of alkaline earth, transition, and heavy metal cations by hydrous oxide gels of iron and aluminum. *Soil Sci. Soc. Am. J.* 40:796–799.

Kivelson, D., and R. Neiman. 1961. ESR studies on the bonding in copper complexes. *J. Chem. Phys.* 35:149–155.

Lakatos, B., L. Korecz, and J. Meisel. 1977a. Comparative study on the Mössbauer parameters of iron humates and polyuronates. *Geoderma* 19:149–157.

Lakatos, B., T. Tibai, and J. Meisel. 1977b. EPR spectra of humic acids and their metal complexes. *Geoderma* 19:319–338.

Lehmann, R.G., H.H. Cheng, and J.B. Harsh. 1987. Oxidation of phenolic acids by soil iron and manganese oxides. *Soil Sci. Soc. Am. J.* 51:352–356.

Lindsay, W.L. 1979. *Chemical equilibria in soils*. Wiley, NY.

Maes, A., and A. Cremers. 1975. Cation-exchange hysteresis in montmorillonite: A pH-dependent effect. *Soil Sci.* 119:198–202.

Manley, E.P., and L.J. Evans. 1986. Dissolution of feldspars by low-molecular-weight aliphatic and aromatic acids. *Soil Sci.* 141:106–112.

Marinsky, J.A., A. Wolf, and K. Bunzl. 1980. The binding of trace amounts of lead(II), copper(II), cadmium(II), zinc(II) and calcium(II) to soil organic matter. *Talanta* 27:461–468.

Martini, G., V. Bassetti, and M.F. Ottaviani. 1980. Mobility and adsorption of copper complexes on aluminas: An ESR study. *J. Chim. Phys.* 77:311–317.

McBride, M.B. 1976. Exchange and hydration properties of Cu^{2+} on mixed-ion $Na^{\pm}$ Cu^{2+} smectites. *Soil Sci. Soc. Am. J.* 40:452–456.

McBride, M.B. 1978a. Retention of Cu^{2+}, Ca^{2+}, Mg^{2+}, and Mn^{2+} by amorphous alumina. *Soil Sci. Soc. Am. J.* 42:27–31.

McBride, M.B. 1978b. Transition metal bonding in humic acid. *Soil Sci.* 126:200–209.

McBride, M.B. 1979a. Chemisorption and precipitation of Mn^{2+} at $CaCO_3$ surfaces. *Soil Sci. Soc. Am. J.* 43:693–698.

McBride, M.B. 1979b. Mobility and reactions of VO^{2+} on hydrated smectite surfaces. *Clays Clay Miner.* 27:91–96.

McBride, M.B. 1980a. Chemisorption of Cd^{2+} on calcite surfaces. *Soil Sci. Soc. Am. J.* 44:26–28.

McBride, M.B. 1980b. A comparative electron spin resonance study of VO^{2+} complexation in synthetic molecules and soil organics. *Soil Sci. Soc. Am. J.* 44:495–499.

McBride, M.B. 1980c. Interpretation of the variability of selectivity coefficients for exchange between ions of unequal charge on smectites. *Clays Clay Miner.* 28:255–261.

McBride, M.B. 1982a. Cu^{2+} adsorption characteristics of aluminum hydroxide and oxyhydroxides. *Clays Clay Miner.* 30:21–28.

McBride, M.B. 1982b. Electron spin resonance investigation of Mn^{2+} complexation in natural and synthetic organics. *Soil Sci. Soc. Am. J.* 46:1137–1143.
McBride, M.B. 1982c. Hydrolysis and dehydration reactions of exchangeable Cu^{2+} on hectorite. *Clays Clay Miner.* 30:200–206.
McBride, M.B. 1985a. Influence of glycine on Cu^{2+} adsorption by microcrystalline gibbsite and boehmite. *Clays Clay Miner.* 33:397–402.
McBride, M.B. 1985b. Sorption of copper(II) on aluminum hydroxide as affected by phosphate. *Soil Sci. Soc. Am. J.* 49:843–846.
McBride, M.B. 1987. Ternary $VO^{2\pm}$ ligand-surface complexes on boehmite and noncrystalline aluminosilicates. *J. Colloid Interfac. Sci.*, 120:419–429.
McBride, M.B., and J.J. Blasiak. 1979. Zinc and copper solubility as a function of pH in an acid soil. *Soil Sci. Soc. Am. J.* 43:866–870.
McBride, M.B., and P.R. Bloom. 1977. Adsorption of aluminum by a smectite: II. An $Al^{3\pm}Ca^{2+}$ exchange model. *Soil Sci. Soc. Am. J.* 41:1073–1077.
McBride, M.B., and D.R. Bouldin. 1984. Long-term reactions of copper(II) in a contaminated calcareous soil. *Soil Sci. Soc. Am. J.* 48:56–59.
McBride, M.B., A.R. Fraser, and W.J. McHardy. 1984. Cu^{2+} interaction with microcrystalline gibbsite. Evidence for oriented chemisorbed copper ions. *Clays Clay Miner.* 32:12–18.
McBride, M.B., B.A. Goodman, J.D. Russell, A.R. Fraser, V.C. Farmer, and D.P.E. Dickson. 1983. Characterization of iron in alkaline EDTA and NH_4OH extracts of podzols. *J. Soil Sci.* 34:825–840.
McBride, M.B., T.J. Pinnavaia, and M.M. Mortland. 1975. Electron spin relaxation and the mobility of manganese(II) exchange ions in smectites. *Am. Mineral.* 60:66–72.
McKenzie, R.M. 1970. The reaction of cobalt with manganese dioxide minerals. *Aust. J. Soil Res.* 8:97–106.
McKenzie, R.M. 1980. The adsorption of lead and other heavy metals on oxides of manganese and iron. *Aust. J. Soil Res.* 18:61–73.
McLaren, R.G., and D.V. Crawford. 1974. Studies on soil copper. III. Isotopically exchangeable copper in soils. *J. Soil Sci.* 25:111–119.
McLaren, R.G., D.M. Lawson, and R.S. Swift. 1986. Sorption and desorption of cobalt by soils and soil components. *J. Soil Sci.* 37:413–426.
McLaren, R.G., R.S. Swift, and J.G. Williams. 1981. The adsorption of copper by soil materials at low equilibrium solution concentrations. *J. Soil Sci.* 32:247–256.
Minnich, M.M., M.B. McBride, and R.L. Chaney. 1987. Copper activity in soil solution: II. Relation to copper accumulation in young snapbeans. *Soil Sci. Soc. Am. J.* 51:573–578.
Miyata, S. 1975. The syntheses of hydrotalcite-like compounds and their structure and physico-chemical properties-I: The systems Mg^{2+}-Al^{3+}-NO_3^-, Mg^{2+}-Al^{3+}-Cl^-, Mg^{2+}-Al^{3+}-ClO_4^-, Ni^{2+}-Al^{3+}-Cl^-, and Zn^{2+}-Al^{3+}-Cl^-. *Clays and Clay Miner.* 23:369–375.
Miyata, S. 1983. Anion-exchange properties of hydrotalcite-like compounds. *Clays Clay Miner.* 31:305–311.
Moraes, J.F.V. 1982. Effect of phosphate on zinc adsorption on aluminum and iron hydrous oxides and in soils. Ph.D. diss., Univ. of California, Riverside.
Mortland, M.M., and M-C. Gastuche. 1962. Cristallisation d'hydroxydes mixtes de magnésium et d'aluminium en milieu dialysé. *Comptes Rend.* 255:2131–2133.
Murad, E., and Schwertmann, U. 1983. The influence of aluminium substitution and crystallinity on the Mössbauer spectrum of goethite. *Clay Miner.* 18:301–312.

Murray, D.J., T.W. Healy, and D.W. Fuerstenau. 1968. The adsorption of aqueous metal on colloidal hydrous manganese oxide. pp. 74–81. In: *Adsorption from aqueous solutions*, W.J. Weber and E. Matijevic (eds.). Advan. in Chem. Ser. no. 79, ACS, Washington, DC.

Murray, J.W. 1975. The interaction of metal ions at the manganese dioxide-solution interface. *Geochim. Cosmochim. Acta* 39:505–519.

Murray, J.W., J.G. Dillard, R. Giovanoli, H. Moers, and W. Stumm. 1985. Oxidation of Mn(II): Initial mineralogy, oxidation state and ageing. *Geochim. Cosmochim. Acta* 49:463–470.

Nalovic, Lj., G. Pedro, and C. Janot. 1975. Demonstration by Mössbauer spectroscopy of the role played by transitional trace elements in the crystallogenesis of iron hydroxides(III). pp. 601–610. In: Proc. Int. Clay Conf., Mexico City.

Newton, D.W. 1971. The influence of pH, phosphate, and silicate on zinc adsorption by clays and soils. Ph.D. diss., Univ. of Illinois, Urbana–Champaign.

Osaki, S., T. Osaki, K. Nishino, and Y. Takashima. 1980. Oxidation and reduction of chromium in natural water. I. Oxidation rate of chromium(III) by oxygen in the presence of manganese(II). *Nippon Kagaku Kaishi* 711–716.

Parks, G.A. 1967. Aqueous surface chemistry of oxides and complex oxide minerals. pp. 121–160. In: *Equilibrium concepts in natural water systems*, W. Stumm (ed.), Advan. in Chem. Ser. no. 67, ACS, Washington, DC.

Perona, M.J., and J.O. Leckie. 1985. Proton stoichiometry for the adsorption of cations on oxide surfaces. *J. Colloid Interfac. Sci.* 106:64–69.

Pingitore, N.E. 1986. Modes of coprecipitation of Ba^{2+} and Sr^{2+} with calcite. pp. 574–586. In: *Geochemical processes at mineral surfaces*, J.A. Davis and K.F. Hayes (eds.). ACS Symposium Series no. 323, ACS, Washington, DC.

Pohlman, A.A., and J.G. McColl. 1986. Kinetics of metal dissolution from forest soils by soluble organic acids. *J. Environ. Qual.* 15:86–92.

Pulford, I.D. 1986. Mechanisms controlling zinc solubility in soils. *J. Soil Sci.* 37:427–438.

Ragland, J.L., and N.T. Coleman. 1960. The hydrolysis of aluminum salts in clay and soil systems. *Soil Sci. Soc. Am. Proc.* 24:457–460.

Rengasamy, P., and J.M. Oades. 1978. Interaction of monomeric and polymeric species of metal ions with clay surfaces. III. Aluminum and chromium. *Aust. J. Soil Res.* 16:53–66.

Russell, J.D., R.L. Parfitt, A.R. Fraser, and V.C. Farmer. 1974. Surface structures of gibbsite, goethite and phosphated goethite. *Nature* (London) 248:220–221.

Sajwan, K.S., and W.L. Lindsay. 1986. Effects of redox on zinc deficiency in paddy rice. *Soil Sci. Soc. Am. J.* 50:1264–1269.

Sanders, J.R. 1982. The effect of pH upon the copper and cupric ion concentrations in soil solutions. *J. Soil Sci.* 33:679–689.

Schindler, P.W., B. Furst, R. Dick, and P.U. Wolf. 1976. Ligand properties of surface silanol groups: I. Surface complex formation with Fe^{3+}, Cu^{2+}, Cd^{2+}, and Pb^{2+}. *J. Colloid Interfac. Sci.* 55:469–475.

Schnitzer, M., and S.I.M. Skinner. 1966. Organo-metallic interactions in soils: 5. Stability constants of Cu^{2+}, Fe^{2+}, and Zn^{2+}-fulvic acid. *Soil Sci.* 102:361–365.

Schnitzer, M., and S.I.M. Skinner. 1967. Organo-metallic interactions in soils: 7. Stability constants of $Pb^{2\pm}$, $Ni^{2\pm}$, $Mn^{2\pm}$, $Co^{2\pm}$, $Ca^{2\pm}$, and $Mg^{2\pm}$ fulvic acid complexes. *Soil Sci.* 103:247–252.

Schoonheydt, R.A. 1982. Ultraviolet and visible light spectroscopy. pp. 163–189. In: *Developments in sedimentology*, vol. 34. *Advanced techniques for clay mineral analysis*. J.J. Fripiat (ed.). Elsevier, NY.

Schwertmann, U., R.W. Fitzpatrick, and J. LeRoux. 1977. Al substitution and differential disorder in soil hematites. *Clays Clay Miner.*, 25:373–374.

Sedlacek, J., E. Gjessing, and J.P. Rambaek. 1987. Isotope exchange between inorganic iron and iron naturally complexed by aquatic humus. *Sci. Total Environ.* 62:275–279.

Senesi, N., G. Sposito, and J.P. Martin. 1986. Copper(II) and iron(III) complexation by soil humic acids: An IR and ESR study. *Sci. Total Environ.* 55:351–362.

Senesi, N., G. Sposito, and J.P. Martin. 1987. Copper(II) and iron(III) complexation by humic acid-like polymers (melanins) from soil fungi. *Sci. Total Environ.* 62:241–252.

Shuman, L.M. 1979. Zinc, manganese, and copper in soil fractions. *Soil Sci.* 127:10–17.

Shuman, L.M. 1986. Effect of ionic strength and anions on zinc adsorption by two soils. *Soil Sci. Soc. Am. J.* 50:1438–1442.

Sidhu, P.S., R.J. Gilkes, and A.M. Posner. 1980. The behavior of Co, Ni, Zn, Cu, Mn and Cr in magnetite during alteration to maghemite and hematite. *Soil Sci. Soc. Am. J.* 44:135–138.

Sposito, G. 1981. *The thermodynamics of soil solutions*. Clarendon Press, Oxford, UK.

Sposito, G. 1986. Distinguishing adsorption from surface precipitation. pp. 217–228. In: *Geochemical processes at mineral surfaces*, J.A. Davis and K.F. Hayes (eds.). ACS Symposium Series no. 323, ACS, Washington, DC.

Stanton, D.A., and R. DuT. Burger. 1970. Studies of zinc in selected Orange Free State soils: 5. Mechanisms for the reaction of zinc with iron and aluminum oxides. *Agrochemophysica* 2:65–76.

Stevenson, F.J., and M.S. Ardakani. 1972. Organic matter reactions involving micronutrients in soils. pp. 79–114. In: *Micronutrients in agriculture*, J.J. Mortvedt et al. (eds.) Soil Sci. Soc. Am., Madison, WI.

Stiers, W., and U. Schwertmann. 1985. Evidence for manganese substitution in synthetic goethite. *Geochim. Cosmochim. Acta* 49:1909–1911.

Stone, A.T. 1986. Adsorption of organic reductants and subsequent electron transfer on metal oxide surfaces. pp. 446–461. In: *Geochemical processes at mineral surfaces*, J.A. Davis and K.F. Hayes (eds.). ACS Symposium Ser. no. 323, ACS, Washington, DC.

Stone, A.T., and J.J. Morgan. 1984a. Reduction and dissolution of manganese(III) and manganese(IV) oxides by organics. 1. Reaction with hydroquinone. *Environ. Sci. Technol.* 18:450–456.

Stone, A.T., and J.J. Morgan. 1984b. Reduction and dissolution of manganese(III) and manganese(IV) oxides by organics. 2. Survey of the reactivity of organics. *Environ. Sci. Technol.* 18:617–624.

Stumm, W., H. Hohl, and F. Dalang. 1976. Interaction of metal ions with hydrous oxide surfaces. *Croat. Chim. Acta* 48:491–504.

Stumm, W., and J.J. Morgan. 1981. *Aquatic chemistry*, 2d ed. Wiley, NY.

Taylor, R.M. 1968. The association of manganese and cobalt in soils—further observations. *J. Soil Sci.* 19:77–80.

Tiller, K.G., J. Gerth, and G. Brümmer. 1984. The relative affinities of Cd, Ni and Zn for different soil clay fractions and goethite. *Geoderma* 34:17–35.

Tiller, K.G., and J.F. Hodgson. 1962. The specific adsorption of Co and Zn by layer silicates. *Clays Clay Miner., Proc. 9th Natl. Conf.* 11:393–403.

Tiller, K.G., V.K. Nayyar, and P.M. Clayton. 1979. Specific and nonspecific sorption of cadmium by soil clays as influenced by zinc and calcium. *Aust. J. Soil. Res.* 17:17–28.

Traina, S.J., and H.E. Doner. 1985a. Copper-manganese(II) exchange on a chemically reduced birnessite. *Soil Sci. Soc. Am. J.* 49:307–313.

Traina, S.J., and H.E. Doner. 1985b. Heavy metal induced releases of manganese(II) from a hydrous manganese dioxide. *Soil Sci. Soc. Am. J.* 49:317–321.

Turner, R.C., and J.E. Brydon. 1965. Factors affecting the solubility of $Al(OH)_3$ precipitated in the presence of montmorillonite. *Soil Sci.* 100:176–181.

Turner, R.C., and J.E. Brydon. 1967. Removal of interlayer aluminum hydroxide from montmorillonite by seeding the suspension with gibbsite. *Soil Sci.* 104:332–335.

Tyler, L.D., and M.B. McBride. 1982. Influence of Cd, pH and humic acid on Cd uptake. *Plant and Soil* 64:259–262.

van Bladel, R., and H. Laudelout. 1967. Apparent irreversibility of ion-exchange reactions in clay suspensions. *Soil Sci.* 104:134–137.

von Zelewsky, A., and J. Bemtgen. 1982. Formation of ternary copper(II) complexes at the surface of silica gel as studied by ESR spectroscopy. *Inorg. Chem.* 21:1771–1777.

Waite, T.D. 1986. Photoredox chemistry of colloidal metal oxides. pp. 426–445. In: *Geochemical processes at mineral surfaces*, J.A. Davis and K.F. Hayes (eds.). ACS Symposium Series no. 323, ACS, Washington, DC.

Waite, T.D., and F.M.M. Morel. 1984. Photoreductive dissolution of colloidal iron oxides in natural waters. *Environ. Sci. Technol.* 18:860–868.

Westall, J.C., and H. Hohl. 1980. A comparison of electrostatic models for the oxide/solution interface. *Adv. Colloid Interfac. Sci.* 12:265–294.

Zinder, B., G. Furrer, and W. Stumm. 1986. The coordination chemistry of weathering: II. Dissolution of Fe(III) oxides. *Geochim. Cosmochim. Acta* 50:1861–1869.

The Components of Nitrogen Availability Assessments in Forest Soils

D. Binkley* and S.C. Hart†

I. Introduction

The availability of nitrogen (N) limits production in many forest ecosystems, and many methods are available for estimating N availability (Keeney, 1980; Binkley 1986; Mahendrappa *et al.*, 1986; Binkley and Vitousek, 1989). The concept of soil availability may represent the rate at which N is converted from unavailable to available forms within the rooting zone (Scarsbrook, 1965). Alternatively, it may refer to the extent to which plant production is constrained by a limited supply of available N. These two aspects of N availability were termed N supply rate and N limitation by Chapin *et al.* (1986). In agroecosystems, N supply rate

*Present address: Department of Forest and Wood Sciences, Colorado State University, Fort Collins, CO 80523, USA.
†Present address: Department of Plant and Soil Biology, University of California, Berkeley, CA 94720, USA.

Advances in Soil Science, Volume 10

and N limitation are often closely linked. In forest ecosystems, differences in species composition, stand age, and soil moisture may uncouple low N supply from N limitation (Chapin *et al.*, 1986). In addition, the nonuniform rooting distribution of trees and the presence of forest floors add spatial complexities to forest N cycles that make it more difficult to estimate N availability in forests than in agroecosystems.

Indexes of soil N availability in forests sometimes covary with each other, but exceptions are common (Lea and Ballard, 1982; Binkley and Matson, 1983; Hart and Binkley, 1985). These indexes have also had mixed success in predicting growth response of forests to N fertilization (Shumway and Atkinson, 1978; Powers, 1980; Lea and Ballard, 1982; Radwan and Shumway, 1983; C. Peterson *et al.*, 1984; Hart *et al.*, 1986). These variable results stem from differences in the relative sensitivity of each method to the factors influencing soil N transformations as well as from differences in the components of the N cycle each method assesses.

In 1980, Keeney reviewed some of the problems in adapting agricultural methods of assessing N availability to forest soils. Since that time, a great deal of information has accumulated, much of it emphasizing the importance of in-field environmental influences and the power of using ^{15}N tracers to identify limiting components of N cycles. In this chapter, we review the major methods used for assessing N availability, emphasizing in-field approaches, the importance of the microbial component, and the uses of ^{15}N tracer methods.

II. The Framework

Over very short time scales (such as 1 day), the N supply rate is regulated by the current pool sizes of inorganic ammonium (NH_4^+) and nitrate (NO_3^-) and the mobility of these ions. On a longer time scale (such as weeks and months), these pools are small relative to the fluxes through them. Differences in fluxes through the inorganic pool result from differences in rates of N mineralization of labile pools or from differences in the rates of immobilization of released inorganic N. The rates of mineralization and immobilization are affected by the chemical composition of the organic pools, the spatial compartmentalization of these pools, and environmental factors such as moisture and temperature. Soil animals may also influence N turnover by increasing the surface-to-volume ratio of detrital material or by partially decomposing the litter before excretion.

The framework depicted in Figure 1 highlights the major factors that influence the amount of N designated as available by N availability indexes. Pools or stocks of N in the soil are denoted by boxes, and flows of N are shown as arrows between the boxes. Environmental influences on soil N transformations (temperature and moisture) are illustrated by the arrow connecting the environment with the soil system. Plant uptake of N, which represents a loss of N from the soil system, is noted as a stock of N outside the soil system but connected via flows from the soil NH_4^+ and NO_3^- pools. The names for the soil organic stocks of N (microbes,

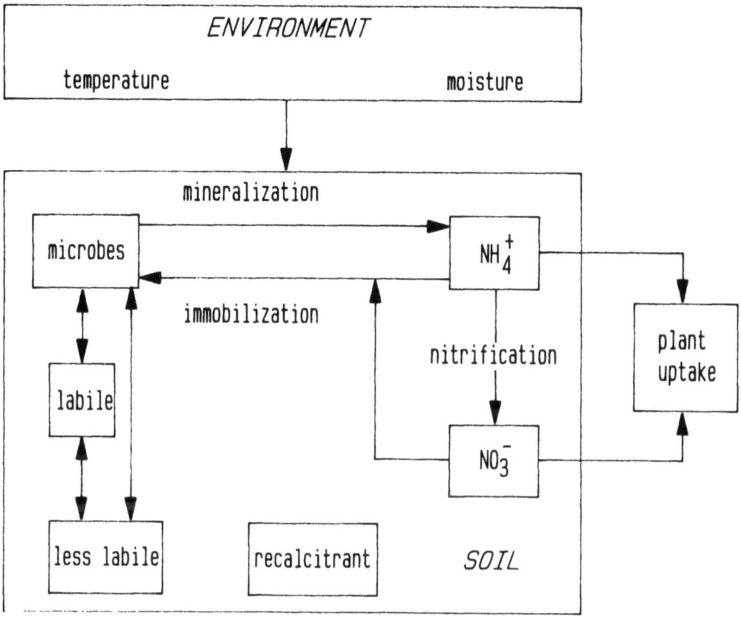

Figure 1. Conceptual framework of the components affecting N availability in forest soils. The relative contribution of each component to estimates of N availability vary among the methods.

labile, less labile, and recalcitrant) are arbitrarily defined; however, they roughly correspond with the microbial biomass, active, stabilized, and old organic matter fractions described by Paul and Juma (1981). The recalcitrant stock of organic N in our framework refers to the pool that has a turnover rate so low relative to the rate of N uptake by trees that it is essentially uncoupled from the inorganic N pools on an annual time scale. This framework will be used to illustrate the various components of the N cycle assessed by each type of N availability method. This will be done by highlighting the stocks or flows a given method assesses as well as by highlighting the environmental factors that influence its estimate of N availability.

III. The Methods

A. Sources of Variability in Indexes of N Availability

The high variability in N availability in forest soils results from vertical horizonation of the soil (especially the forest floor and mineral soil divisions), small-scale differences in soils owing to the effects of microtopography and individual trees, and larger-scale effects owing to geomorphologic variation across landscapes. In addition to spatial sources of variability, handling and storage of samples can

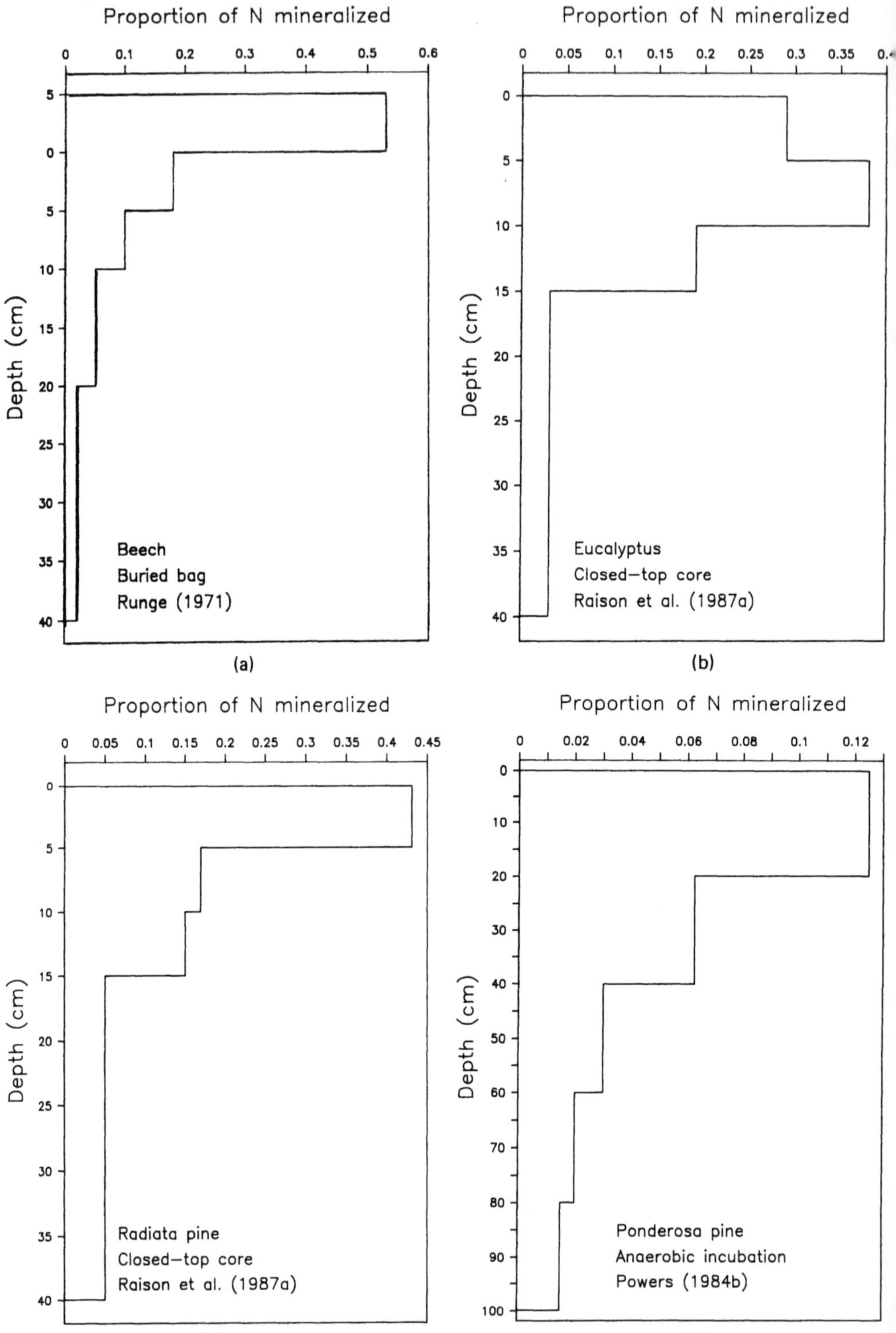

Figure 2. Depth profiles of N availability, expressed as proportions of the total amount mineralized in each 5-cm-depth interval of the profile. Substantial mineralization occurs

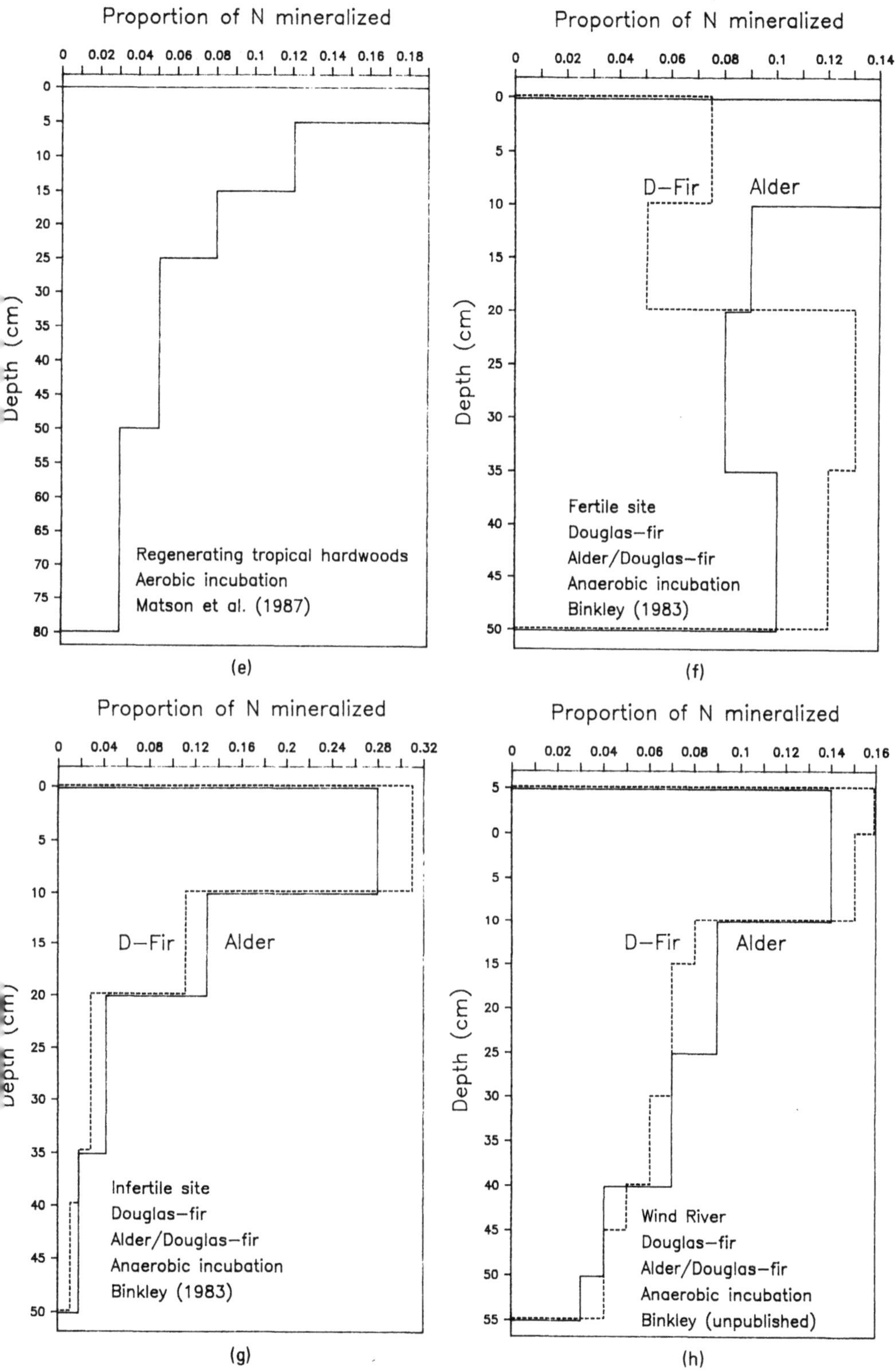

at depths below the 15-cm depth normally sampled, and the presence of N-fixing alders may change the depth profile as well as the magnitude of N availability.

introduce large variations in estimates of N availability. Many of the assessment methods involve incubations for defined time periods; the choice of time period has a large effect on availability estimates.

1. Spatial

Most studies of N availability examine only the upper mineral soil (0–10 or 0–15 cm), and occasionally the forest floor (see later discussion). The forest floor and 0- to 15-cm depth mineral soil typically produces half or more of the N mineralized in forests, and mineralization generally decreases with depth (Figure 2). Major exceptions occur; for example, mineralization in the 5- to 10-cm depth in a *Eucalyptus* ecosystem exceeded the rate in the 0- to 5-cm depth (Raison *et al.*, 1987a). In a Douglas-fir [*Pseudotsuga menziesii* (Mirb.) FRANCO] plantation on a very fertile site, the N mineralized during anaerobic incubations actually increased with depth (Binkley, 1983). The pattern of N mineralization with depth depends, in part, on the distribution of total N, which generally decreases with soil depth. However, Federer (1983) found that N mineralized per g of organic matter was not constant with depth and that variations in this ratio followed no consistent pattern across the sites he examined.

The degree of horizontal variability in N availability depends, in part, on the method of assessment. The coefficients of variation (standard deviation as a proportion of the mean) are especially high for pools of inorganic N and for ion exchange resin bags, typically averaging from about 0.8 to > 1.0 (Table 1). This variability is less handicapping for ion exchange resin bags because differences between sites or treatments are generally very large, whereas inorganic N pools typically differ less among sites or treatments. In contrast, total soil N shows much less variability, with coefficients of variation of about 0.2.

Spatial statistics (geostatistics) can take advantage of variability among samples within sites or treatments and provide insights about the spatial scale of variation and about efficient sampling designs (see R. Webster, 1985). For example, Robertson (1987) characterized the spatial variability of inorganic nitrogen in a 0.5-ha plot in an abandoned agricultural field (old-field) in Michigan. Samples were collected in a grid at 4.6-m intervals. He found that samples taken < 20-m apart were highly correlated (Figure 3); an efficient sampling design for estimating the mean value for a large plot would involve a minimum distance between subsamples of about 40 m. The grid sampling design also allowed isopleths to be calculated for the plot (Figure 4), illustrating the spatial variability of inorganic N (see also Robertson *et al.*, 1988).

2. Temporal

Repeated samplings of total soil N through a single year should produce essentially the same estimate because the annual fluctuations in total soil N (in the absence of fertilization) are very small relative to the total pool. The pool size of inorganic N shows strong seasonal dynamics because it is small relative to annual fluxes. The ratio of seasonal maximum to minimum for inorganic N is commonly

Table 1. Within-plot variability in methods of assessing nitrogen availability, for uncomposited samples only

Method and reference		Coefficient of variation	Seasonal maximum / Seasonal minimum
Extractable ammonium and nitrate			
Christensen and MacAller 1985	NH_4		2
	NO_3		4
Covington and Sackett 1986	NH_4	0.3	5
	NO_3	1.3	13
Frazer et al. in review	NH_4	0.5	
	NO_3	1.1	
Glavac and Koenies 1978a	Sum		10
Gosz and White 1986	NH_4		5
	NO_3		10
Hart and Binkley 1985	NH_4	0.4	
	NO_3	0.2	
Mladenoff 1987	NH_4		2–8
	NO_3		8–10
Montagnini *et al.* 1986	NH_4	0.6	3
	NO_3	1.0	6
Runge 1974	Sum		5
Ryan and Covington 1986	NH_4		5
	NO_3		7
Strader *et al.* 1988	NH_4	0.8	1.8
	NO_3	0.9	2.0
Virzo De Santo *et al.* 1982	NH_4		2
	NO_3		8
	Sum		5
Vitousek and Matson 1985	NH_4		4
	NO_3		>5
Boiling water/salt			
Binkley and Matson 1983		0.3	
Hart and Binkley 1985		0.4	
Autoclaving			
Binkley and Matson 1983		0.3	
Total N			
Binkley 1982		0.4	
Binkley and Matson 1983		0.3	
Binkley *et al.* 1982		0.3	
Fox *et al.* 1986		0.2	
Matson and Boone 1984		0.1	

Table 1. *Continued*

Method and reference		Coefficient of variation	Seasonal maximum / Seasonal minimum
Anaerobic incubations			
Adams and Attiwill 1986b		0.16	2
Binkley 1982		0.75	
Binkley and Matson 1983		0.52	
Binkley *et al.* 1982		0.32	
Hart and Binkley 1985		0.23	
Powers 1980, 1984b		0.45	2
Aerobic incubations			
Binkley and Matson 1983	NH_4	0.4	
	NO_3	1.1	
	Sum	0.6	
Hart and Binkley 1985	NH_4	0.6	
	NO_3	0.2	
Matson and Vitousek 1981	NO_3	0.6	
	Sum	0.5	
Montagnini *et al.* 1986	NO_3	0.8	1.8
	Sum	0.7	1.4
Olson and Reiners 1983	NH_4	0.1	1.5
	NO_3	0.1	1.4
	Sum	0.1	1.3
Adams and Attiwill 1986b	Sum		2.3
Raison *et al.* 1987a	NH_4	0.2	
	NO_3	<0.1	
	Sum	<0.1	
Thorne and Hamburg 1985	NO_3	0.3	
Virzo De Santo *et al.* 1982	NO_3		8
	Sum		8
Vitousek and Matson 1985	NO_3		>10
	Sum		>10
Ion exchange resin bags			
Binkley and Matson 1983	NH_4	0.8	
	NO_3	1.5	
Binkley *et al.* 1986	NH_4	0.8	
	NO_3	0.6	
	Sum	0.6	
Hart, unpublished	NH_4	0.7–1.1	14
	NO_3	0.9	55–95
	Sum	0.8–1.1	40–45
Hart and Binkley 1985	NH_4	1.3	
	NO_3	0.8	
Buried-bag incubations			
Gordon and Van Cleve 1983	NH_4	0.7	
Hart and Firestone in review	Sum	0.3–0.6	8–17

Table 1. *Continued*

Method and reference		Coefficient of variation	Seasonal maximum / Seasonal minimum
Buried-bag incubations cont.			
Matson and Boone 1984	Sum	0.6	8
Mladenoff 1987	NO_3		3
	Sum		4
Nadelhoffer *et al.* 1983	NO_3		8
	Sum		16
Runge 1971, 1974	Sum	0.2	3
Poovarodom *et al.* 1988	Sum	0.4–0.7	15
Closed-top core incubations			
Van Praag and Weissen 1973	Sum	0.7	13
Raison *et al.* 1987a	NH_4	1.6	
	NO_3	3.0	
	Sum	1.6	
Resin-core incubations			
Bloss and Binkley 1989	Sum	0.3	
Hart and Firestone 1989	Sum	0.4–0.5	6–9
Powers 1987	Sum	0.1–0.3	
Sasser and Binkley 1988	Sum	0.4	
Strader et al. 1988	Sum	0.3–0.5	2–

fivefold or greater (Table 1). Even methods that are designed to index the potentially mineralizable N (such as aerobic and anaerobic incubations under laboratory conditions) often show a twofold seasonal variation.

3. Handling and Storage

The handling and storage of forest soils after sampling can affect N availability estimates and the interpretations derived from them. Air-drying and sieving are common practices because refrigeration of soil samples is costly and homogenization of the sample reduces variability among subsamples. Air-drying and grinding soils usually increases extractable inorganic N pools (see Gilliam and Richter, 1985). The pools may also increase with the length of storage after air-drying (see, Powers, 1984c). Air-drying may initially decrease or increase mineralizable-N, as measured by the increase in NH_4^+-N after anaerobic incubation, compared to field-moist soils (McNabb, 1984). However, mineralizable-N does not seem to change further after air-drying and storing for periods of up to a year (Powers, 1984c). Hart and Binkley (1985) found that air-drying several forest soils prior to anaerobic incubation increased N mineralization by about 70% over field-moist soils, but values for air-dried soils correlated well (r^2 = 0.8) with field-moist values.

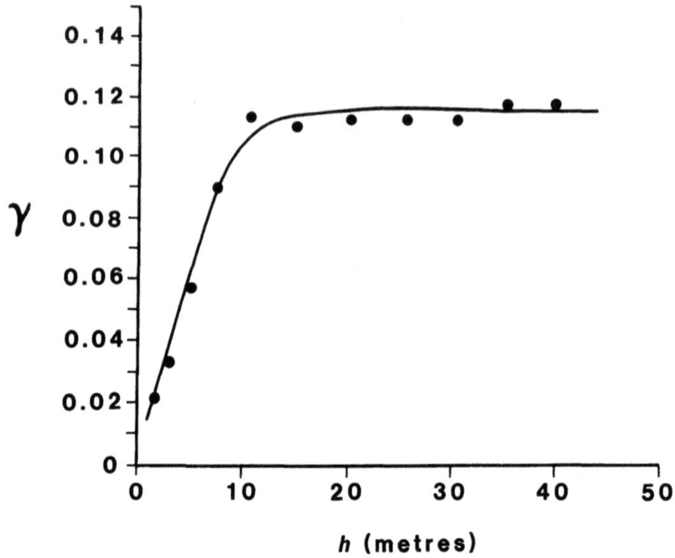

Figure 3. Semivariogram of ammonium + nitrate in an old-field in Michigan. Sample points closer than 20 m covary strongly; a distance of 40 m between samples is needed for complete independence. (Robertson, 1987; reprinted by permission of the Ecological Society of America.)

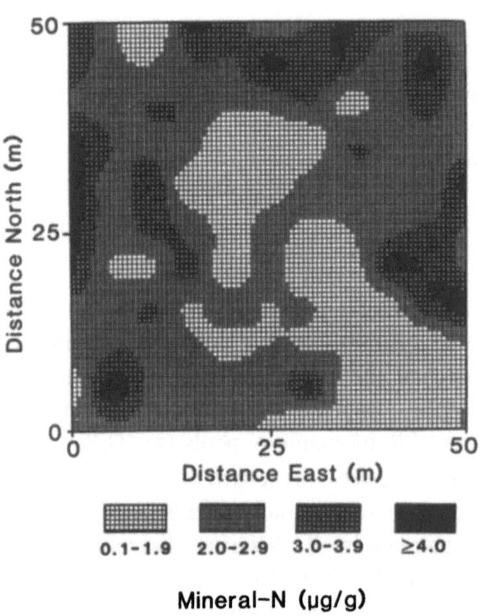

Figure 4. Isopleths of ammonium + nitrate in an old-field in Michigan, based on a kriging routine. (Robertson, 1987; reprinted by permission of the Ecological Society of America.)

Air-drying also affects rates of N production under aerobic conditions. Birch (1958, 1960) documented that rewetting air-dried soils produced higher rates of N mineralization than found for soils that were continuously moist. He also showed that successive drying and wetting cycles produced similar amounts of N. He suggested that this phenomenon resulted from the high metabolic activity of the recolonizing microbial population found after air-dry soils are rewetted. Since Birch's pioneering work, other researchers have found similar trends (e.g., Van Schreven, 1967; Marumoto *et al.*, 1977a, 1977b; Sparling and Ross, 1988), including for some forest soils (Turner and Franz, 1985). Sahrawat (1980) found that rewetting air-dried soils enhanced net NH_4^+ production but retarded net nitrification. Nordmeyer and Richter (1985) demonstrated rewetting air-dried soils stimulated N mineralization in soils using the repeated incubation technique; the effect was short-lived, lasting only 10 to 20 days after rewetting.

Air-drying kills much of the microbial biomass, and enhanced N mineralization probably results from the breakdown of this readily mineralized substrate by surviving microbes (Seneviratne and Wild, 1985; Sparling and Ross, 1988). For example, Sparling and Ross (1988) found a decrease in the microbial biomass-N pool after soil drying, along with a narrowing of the ratio of net C mineralized to net N mineralized during incubation.

Soil sieving breaks down soil aggregation and reduces the spatial compartmentalization between active microorganisms and labile substrates. Thus, it is not surprising that soil sieving can greatly influence biological measures of N availability. Satterson (1985) found that sieving doubled the rate of net mineralization in mineral soils but had no effect on forest floor materials. Raison and others (1987a) showed that sieved soils (< 5 mm) mineralized two- to ten-times more N than undisturbed soils during incubation in the field; sieving also increased nitrification. Nordmeyer and Richter (1985) noted that soil sieving caused an initial flush in N mineralization compared to undisturbed soils using the repeated incubation technique. This flush was similar to that found with air-drying.

Soil storage also can significantly influence N availability estimates. Soil stored in an air-dry state may show variable N availability assay values for a short period following sampling and storage; this variability generally declines after a few weeks (McNabb, 1984). However, storage of field-moist soil (frozen or at 2 to 5 C) is more difficult. Freezing may release moderate amounts of NH_4^+ (Hart and Firestone, 1989), possibly as a result of the lysing of microbial cells. Nevertheless, the amount of N released for a given soil after freezing appears similar for soils sampled at different times. Thus estimates of N production where preincubation and postincubation soils have both been frozen are likely to be unaffected by this method of storage (Hart and Firestone, 1989).

Although rates of N mineralization and nitrification are reduced at low temperatures, significant accumulations of NH_4^+ and NO_3^- can occur over fairly short periods of cold storage. Hart (unpublished data) found that NO_3-N and total inorganic N soil pools increased between 6 to 23% and 1 to 19%, respectively, for some forest soils after only 1-week storage at 5 C. Thus, the length of sample storage at 2 to 5 C needs to be minimized or taken into account when estimating N production rates from cold-stored soils.

Ideally, extractions of soils for estimating initial inorganic N pools sizes should be made in the field using preweighed containers with a known amount of extractant. Rocks and roots can be removed by hand prior to extraction or accounted for later back in the laboratory by washing or sieving. If field extraction is not practical, field-moist soils should be kept cold on ice extracted in the laboratory.

4. Incubation Period

In methods that involve incubation of soil samples, the period of incubation has a large effect on the estimate of N availability. Laboratory methods typically use an arbitrary, short time period (such as 7 to 30 days), or they use several-month periods coupled with periodic leaching to remove inorganic N. The net production of inorganic N is typically not a linear function of the length of incubation, owing to changes in the balance between mineralization and immobilization and to the dynamics of microbial populations. The sensitivity of net mineralization values to incubation period is also important for in-field techniques. For example, a sequence of four 1-month incubations in buried bags in an alder/conifer ecosystem in Oregon produced about 30% more NH_4^+ plus NO_3^- than did 2-month incubations (Figure 5). A single 4-month incubation produced only about half the amount found in the four 1-month incubations. We know of no study that has characterized the components of these differences for in-field incubations, but ^{15}N techniques offer substantial opportunities.

B. Chemical Indexes

1. Extractable Ammonium and Nitrate

Ammonium and NO_3^- are the forms of nitrogen that supply the majority of plant uptake, and the quantity of these ions from soil samples has often been used as an index of N availability. The basic procedure (Bremner, 1965; Keeney and Nelson, 1982) involves extraction of soil samples for 1 hour with 2 *M* KCl in a ratio of 10 mL solution/1 g mineral soil. Filtered or centrifuged extracts can be stored for several months at low temperature with no change in chemistry. These extraction procedures also apply to methods discussed later.

The levels of extractable NH_4^+ and NO_3^- in soils depend strongly on all the components in Figure 1. These pools are often very dramatic, with subsamples from single plots showing high variability both spatially and temporally. The coefficients of variation among subsamples from one plot are typically about 0.3 to 1.0 for extractable NH_4^+ plus NO_3^- (Table 1). Through a season, the mean values for extractable NH_4^+ and NO_3^- often vary by twofold to tenfold.

Because so many factors affect the amount of NH_4^+ and NO_3^- that can be extracted from a soil sample at any given time, these pools typically are not capable of providing reliable information on N availability. Recent use of these pools has been limited primarily to studies of potential allelopathic effects of vegetation and the effects of fires.

Rice and Pancholy (1972) measured pools of extractable NO_3^- in a successional sequence of ecosystems and concluded that the decrease in extractable NO_3^- with

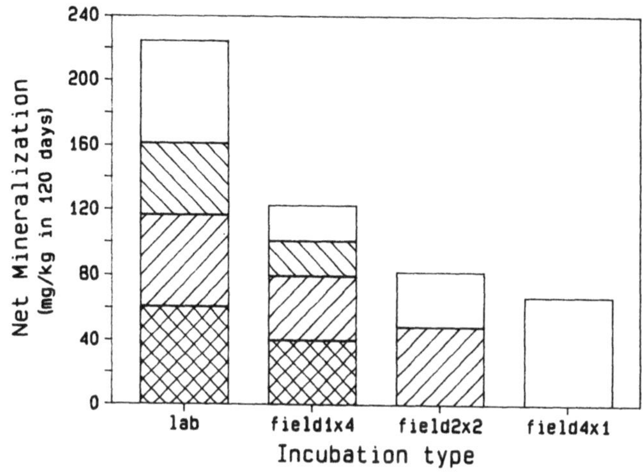

Figure 5. Mineralization in intact soil cores over a 4-month period (fresh samples used for each period): laboratory = four 1-month incubations; field 1 × 4 = four 1-month buried-bag incubations; field 2 × 2 = two 2-month buried-bag incubations; and field 4 × 1 = one 4-month incubation. The laboratory rate for the four 1-month incubations was about 90% greater than the buried-bag rate. The rate of mineralization declined for the longer periods of incubation (Binkley, unpublished data).

successional development demonstrated decreased nitrification. However, Robertson and Vitousek (1981) argued that extractable pools were poor indexes of the fluxes of NH_4^+ and NO_3^-. They showed that extractable NO_3^- did not relate well to NO_3^- produced during incubations. Indeed, they found that a late-successional oak site had the highest level of extractable NO_3^- among the ecosystems they studied, but it also showed the lowest rate of net nitrification. In a similar study, Christensen and MacAller (1985) found that an old-field site had higher soil NO_3^- in May than nearby forested soils but that no differences were found in August nor were differences found between sites using aerobic incubations.

Extractable pools have provided more interpretable data on N availability in some studies that examined the effects of fire. Ryan and Covington (1986) measured extractable NH_4^+ in the early summer in a ponderosa pine (*Pinus ponderosa* Laws.) forest that had been prescribe burned the previous autumn. They found that NH_4^+ concentrations on burned plots were up to 80 times greater than on control plots, but that NO_3^- concentrations on burned plots were only about twice as great as on unburned plots. In this case, the treatment effect on N availability was so large (the increase in extractable NH_4^+ was 70 kg N/ha) that possible short-term fluctuations in extractable N pools were less handicapping than in the alleopathy example above. In Israel, Kutiel and Naveh (1987) found that a wildfire in an Aleppo pine (*Pinus halepensis* Mill.) forest doubled soil NH_4^+ pools and increased NO_3^- pools by tenfold. They verified the increase in N availability in a bioassay with wheat (*Triticum aestivum*); plants in the burned soils accumulated three times as much N as those in control soils. Rashid (1987) examined the

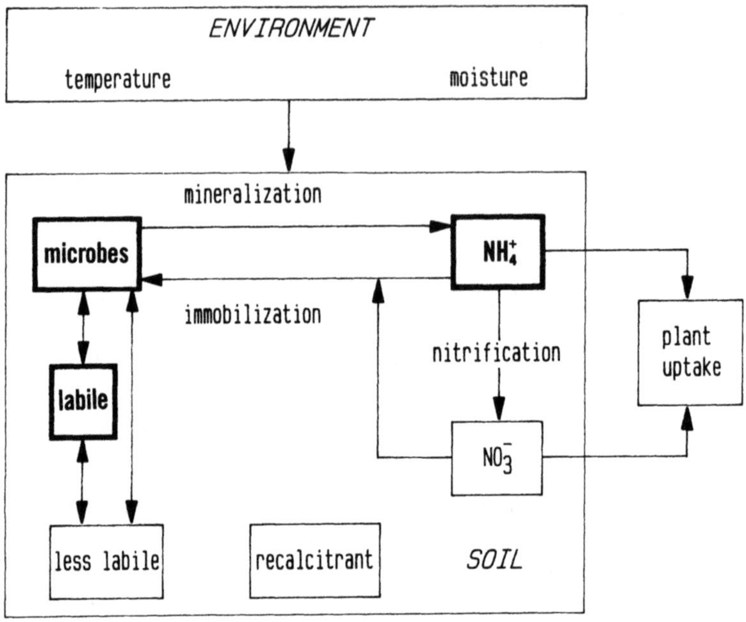

Figure 6. Chemical extractions of soil samples include varying importances of the ammonium, microbial, and labile components, but microbial mineralization and immobilization are not involved.

effects of a wildfire in an oak forest in Algeria and found that NH_4^+ concentrations were increased fivefold in the burned site but that NO_3^- concentrations showed no major difference. Aerobic incubations verified that mineralizable-N was marginally greater in burned soils.

2. Chemical Extractions of Labile Pools

The nitrogen mineralized by microbes is thought to come primarily from an active phase of soil organic matter (Figure 6) that consists of compounds that are readily degraded by microbial enzymes (see Paul and Juma, 1981). A variety of chemical techniques have been developed that attempt to extract this phase directly, thus bypassing the need to use microbes. Approaches used on agricultural soils include boiling samples in water (Keeney and Bremner, 1966) and in KCl (Gianello and Bremner, 1986); autoclaving samples (Stanford and DeMar, 1969); extracting samples with various acidic and alkaline compounds (see Keeney, 1982); and electro-ultrafiltration (Németh *et al.*, 1979). Following extraction, analysis may include either the NHl_4^+ released, the distillable N (NH_4^+ plus any compounds volatile in alkaline solutions), or the total soluble N.

In forest soils, most attention has focused on the boiling water/salt and on the autoclaving methods. Mahendrappa (1980) compared the N distilled from autoclaved samples with the NH_4^+ extracted from incubated samples and found a good correlation ($r^2 = 0.6$). S. Webster (1984) found that the NH_4^+ released by boiling water and by autoclaving correlated well (r^2 of about 0.9) with total N in soils from 14 Douglas-fir [*Pseudotsuga menziesii* (MIRB.) Franco] stands. He found that boiling water released about 4% of the total N as NH_4^+ and the autoclaving released about 6% as NH_4^+. The total soluble N produced from autoclaving contained about 13% of the total N in the soil samples. Webster also reported that both methods also correlated with N released during anaerobic incubations (r^2 = 0.6). Binkley and Matson (1983) compared a variety of methods on 9 replicate soils from each of 3 stands (with and without N-fixing *Alnus*). They found that a high correlation between the boiling water and autoclave methods ($r^2 = 0.9$) and both methods correlated well with total N ($r^2 = 0.8$). Binkley and Matson also compared the NH_4^+ released by these treatments with the total soluble N produced. They found that although the total soluble N was about 4 times the NH_4^+ released, the two N forms correlated highly ($r^2 = 0.9$). Therefore, the added digestion step necessary to determine total soluble N did not appear warranted. Lea and Ballard (1982) found that the inorganic N released by autoclaving of 31 soils correlated moderately with incubation indexes (r^2 = 0.3 to 0.5) but not with stand response to N fertilization. Hart and Binkley (1985) examined correlations among several indexes of N availability in 11 stands of loblolly pine (*Pinus taeda* L.). They found that boiling salt-extractable NH_4^+ did not correlate well with any other measure of N availability in either fertilized or unfertilized soils and that the method could not predict tree-growth response to fertilization (Hart *et al.*, 1986).

Acidic and alkaline extractions have not received much use in forest soils, but they may have some potential. For example, Fisher and Stone (1969) used a hydrofluoric acid extraction to examine patterns of N availability within 5 conifer plantations. They found that HF-extractable N near trees represented a greater proportion of the total N (in all 5 stands) than it did in between-tree portions of the plantations.

These chemical extraction methods (especially the boiling water/salt method) are generally easier than incubation methods and have often proven as successful at predicting nutrient limitation in forests as other methods. This approach may be particularly useful where high rates of microbial immobilization obscure differences in mineralization rates and N availability to trees. Future studies should include one or more of these chemical extraction methods, and focus particularly on the opportunities for comparing the components underlying availability estimates using ^{15}N methods.

3. Total N

Two general methods are used for determining the total N content of soil samples (Bremner and Mulvaney, 1982). The Kjeldahl approach involves wet digestion of samples with a strong acid, followed by determination of the NH_4^+ in the digest. The Dumas approach involves dry combustion and determination of the N_2 produced (more oxidized forms are first reduced to N_2).

Studies that have examined relationships between total N and measures of N availability have often found good correlations, with r^2 typically ranging from about 0.5 to 0.9 (Powers, 1980; Lea and Ballard, 1982; Binkley and Matson, 1983; Radwan and Shumway, 1983; S. Webster, 1984). In fact, predictions of Douglas-fir response to N fertilization based on total N are often not significantly poorer than those based on anaerobic incubations (Powers, 1980; Lea and Ballard, 1982; Radwan and Shumway, 1983; Shumway, 1984). Correlations between total soil N and fertilization response are generally negative in the Douglas-fir region, indicating that high responses generally occur on sites low in total N.

The availability of N to trees depends, in part, on competition with heterotrophs and the ratio of total C/total N appears in some cases to be a better index of N availability than total N alone (see C. Peterson *et al.*, 1984). Because not all C is available to microbes, the ratio of available C/total N may be superior to total C/total N (see Zöttl, 1960; Flanagan and Van Cleve, 1983; Harmer and Alexander, 1986).

These studies generally support the conclusion that the percentage of the total N measured by laboratory availability indexes is less variable than is the total N content across sites, although exceptions occur (see Strader *et al.*, 1988). When the range of total N contents spans a large range (such as an order of magnitude), the differences in substrate quality and microenvironment appear less important. Therefore, it seems valid to conclude that at a regional scale N availability may be indexed adequately by total N (see also Post *et al.*, 1985). At a local scale, where sites may not differ greatly in total N, differences in substrate quality and microenvironment may be important and in-field incubations may provide better indexes of N availability.

C. Laboratory Incubations

1. Aerobic Incubations

One of the most frequently used methods of N availability assessment is aerobic incubation under controlled environmental conditions. These assays are relatively easy, they mimic field conditions and provide an estimate of the importance of nitrification (Figure 7). The technique typically involves incubating a sieved soil sample under moist (near field capacity) conditions at 20 to 25 C for 10 to 30 days. Various techniques are used to keep soil water content constant during the incubation period, such as adding water to replace evaporated water at frequent intervals; using closed containers with small holes at the top to reduce evaporation and maintain gas exchange; covering containers with thin plastic film that is readily permeable to O_2 and CO_2 but less permeable to water vapor; and carrying out the incubation in containers with a large headspace that is purged frequently with water-sataurated air. Both NH_4^+ and NO_3^- are measured in salt solution extracts (such as 2 *M* KCl) of subsamples before and after incuba-

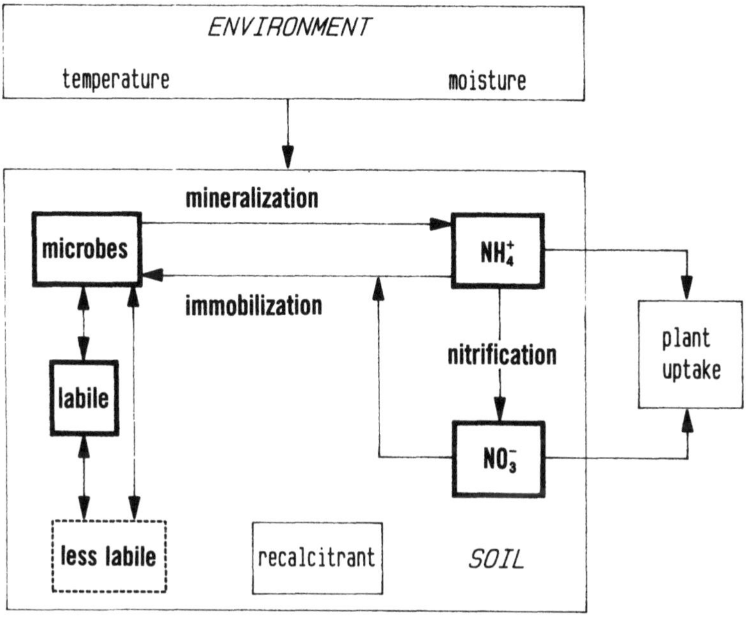

Figure 7. Estimates of N availability from aerobic incubations depend on the nitrate, ammonium, microbial and labile components; repeated incubations may also be affected by the less labile component. Microbial activities are fundamental to these methods, but method-induced changes in these activities can be problematic.

tion. The net change in NH_4^+ + NO_3^- concentrations provides an estimate of net mineralization. The changes in individual N forms are sometimes reported, with the net change in the NH_4^+ pool referred to as net ammonification and the net change in the NO_3^- pool as net nitrification.

The length of the incubation can greatly affect the N availability estimate (Johnson *et al.*, 1980). Recent disturbances of soil samples (such as sieving and rewetting) markedly alter N transformations (Runge, 1974; Nordmeyer and Richter, 1985). Because of this disturbance factor, longer incubation periods are preferred (3+ weeks). Aerobic incubations lasting 1 month assess primarily the partial turnover of the microbial biomass and labile soil organic matter pools (Figure 7). Longer aerobic laboratory incubations (90 to 160+ days) have been used to determine the size of the entire active fraction of soil organic matter (Paul and Juma, 1981), which is roughly comparable to the labile + microbial soil organic matter pools in our framework.

Some forest soils that apparently do not have a very high capacity to nitrify based on 1-month incubations may show substantial accumulations of NO_3^- when the incubation period is lengthened (Johnson *et al.*, 1980; Vitousek *et al.*, 1982; Gosz and White, 1986). Thus, the relative importance of NO_3^- compared to NH_4^+ in aerobic incubations depends strongly on the chosen length of incubation.

Modification of the aerobic incubation technique to assess CO_2-C production and N mineralization simultaneously can provide valuable insight regarding the nature of the organic substrates being mineralized (Johnson *et al.*, 1980; Hart, 1988; Sparling and Ross, 1988).

The moisture content of soils during incubation significantly affects net N mineralization and nitrification rate estimates with the aerobic incubation method (Stanford and Epstein, 1974). Matson and Vitousek (1981) found that the rate of net N mineralization and nitrification doubled when the moisture content of soils from cut and uncut hardwood forests was increased from 15 to 35% of the sample weight. Further, the relationship between soil moisture and net N production was not always similar among the soils. Because of this strong and variable moisture influence, some researchers prefer to use field-moist soil samples for aerobic incubations (Hart and Binkley, 1985; Hart, 1988). Use of field-moist soils may provide better estimates of actual mineralization occurring in a site at a given time, but such estimates may show much greater variability across sampling periods than would constant-moisture incubations. Finally, it should be noted that aerobic incubation estimates of N production conducted in the laboratory may or may not reflect field estimates (Lamb, 1980; Matson and Vitousek, 1981; Hart and Binkley, 1985; Adams and Attiwill, 1986b). For example, the alder/conifer soil in Figure 2 showed twice the rate of net mineralization during four 1-month aerobic incubations in the laboratory compared to in-field incubations.

2. Anaerobic Incubation

Another simple laboratory-based method of N availability assessment involves incubation of soil samples under water in closed containers at 30 or 40 C for 14 or 7 days (S.A. Waring and Bremner, 1964; Keeney, 1982). Several variations of this method include purging the headspace of the extraction container with N_2 to lower the redox potential, and measuring CO_2-C production as well as N production. The term *anaerobic incubation* is commonly used, although actual measurements of the redox potential of the soil is not routinely included. Some of the problems associated with the aerobic method are alleviated with the anaerobic technique, including the problem of maintaining water content during aerobic incubation. The anaerobic laboratory incubation is sensitive to only a few components of the N cycle (Figure 8). In this method, only the net change in the NH_4^+ pool is assessed because any NO_3^- present is presumably denitrified and no new NO_3^- should be formed. Nitrification is not assessed by the anaerobic incubation method. The greater mobility (Barber, 1974; Binkley, 1984b) as well as the smaller degree of immobilization of NO_3^- relative to NH_4^+ (Jansson, 1958; Shen *et al.*, 1984; Nishio et al., 1985) in soils suggests that nitrification may be especially important for tree nutrition. Therefore, this method may not be useful in comparing N availability among sites where nitrification rates are variable and important.

The NH_4^+ flush that occurs during anaerobic incubation has been thought to come from the decomposition of obligate aerobic microorganisms by anaerobic

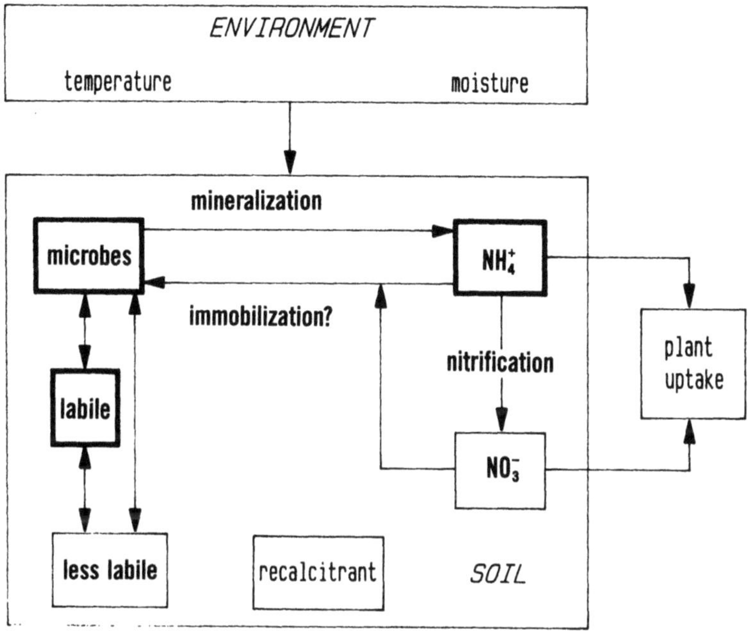

Figure 8. Anaerobic incubations include the ammonium, microbial, and labile components. Microbial activity is important, but the unusual conditions of the incubations result in rates and patterns that do not mimic in-field situations.

bacteria (Adams and Attiwill, 1986b; Paul *et al.*, 1986; Myrold, 1987). In contrast, a recent study using added ^{14}C- and ^{15}N-labeled microorganisms has shown that less than 25% of the N mineralized during anaerobic incubation at 25 C came from microbial biomass (Paul *et al.*, 1986). However, other studies have shown that N mineralized during anaerobic incubation does originate primarily from the soil microbial biomass (Myrold, 1987). Myrold suggested that the greater proportion of the microbial biomass released by anaerobic incubation in his study (33 to 78%) compared to that of Paul *et al.* (1986) was due to the higher incubation temperature (40 C) used by Myrold (1987). Further evidence for the presumption that anaerobic incubation measures primarily microbial biomass is provided by the strong correlation ($r^2 > 0.9$) between the amount of CO_2 evolved during anaerobic incubation and an estimate of microbial C for several forest soils (Smith *et al.*, 1985).

As with aerobic incubations, modification of the standard anaerobic technique to simultaneously assess C and N mineralization may aid in understanding and interpreting variable anaerobic assay results (Paul *et al.*, 1986; Smith *et al.*, 1985). For example, analysis of the net ratios of C/N mineralized may provide insight into which soil organic matter components are being decomposed.

Sollins *et al.* (1984) used anaerobic incubations to examine the role of organic matter from light and heavy density fractions. The light fraction from six sites consisted of partly decomposed root fragments and other recent organic matter, and the heavy fraction consisted of organic matter adsorbed or deposited onto mineral surfaces. For five of the six soils, the heavy fraction mineralized more N during anaerobic incubations than was mineralized from the light fraction. In addition, the net mineralization from the heavy fraction exceeded the net rate for incubations of whole soil, probably owing to immobilization of N within the light fraction. Future studies that combine fractionation techniques with isotope methods could further clarify the importance of various fractions as sources and sinks of available N.

In some forest soils, N mineralization under anaerobic conditions is less than that under aerobic incubations for similar time periods and temperatures (Smith *et al.*, 1981). The opposite pattern has also been found for anaerobic and aerobic incubations conducted in the field (Adams and Attiwill, 1986b; Powers, 1987) and for some agricultural soils (Smith *et al.*, 1981). Reduced N immobilization under anaerobic conditions may explain the higher N production estimates found for some forest soils compared to aerobic values. Anaerobically mineralized N estimates may also be more sensitive to the relative sizes of labile- and microbial-N than aerobic incubation methods. Net NH_4^+ production for 7-day anaerobic incubations at 40 C is usually greater than net mineralization in 1-month aerobic incubations at 20 C (Binkley and Matson, 1983; Hart and Binkley, 1985). Nevertheless, the methods correlate highly with each other ($r^2 = 0.8$ to 0.9) in most soils, with notable exceptions (see Hart and Binkley, 1985).

Powers (1980) compared anaerobic incubations in the laboratory and in the field along an elevational sequence of sites. The ratio of N produced during 1-year anaerobic incubations in the field to the laboratory rates (at 30 C) increased with increasing mean soil temperature in the field (Powers, 1984b). Ratios rose by about two-thirds for each 10 C rise in soil temperature. These results emphasize the importance of incubation temperature in biological assays.

Geist (1977) found that the anaerobic index correlated well with N uptake by orchard grass (*Dactylis glomerata* L.) tops or top yield for some volcanic ash forest soils in Oregon, whereas an aerobic incubation did not.

Adams and Attiwill (1986b) found for some eucalyptus sites in Australia that changes in the amount of N mineralized during anaerobic incubations in the laboratory tended to follow closely the fluctuations in the pattern of net mineralization that occurred during aerobic incubation of the soils in the field.

3. Repeated Incubations

Stanford and Smith (1972) and Stanford *et al.* (1974) developed a method in which repeated incubations of soil samples are used to derive an estimate of the portion of soil N that is potentially mineralizable and the steady-state rate of N release. This procedure involves aerobic incubation of samples in containers that can be leached with dilute salt solutions to remove accumulated NH_4^+ and NO_3^-. Samples may be mixed with sand to improve drainage. The samples are incubated for several months, with salt rinses twice a month. In some cases, non-N nutrient solutions are added after each rinse. These authors assume that mineralization follows first-order kinetics, and they calculate a "nitrogen mineralization potential" (N_0) that represents the amount of potentially mineralizable N and a constant (k) that corresponds to the proportion of the remaining N that is mineralized in each time period. Although mineralization rates often resemble first-order kinetics, the interacting processes that regulate mineralization and immobilization reactions of the diverse array of organic compounds in forest soils are not mechanistically constrained to first-order properties. Other studies indicate that second-order or higher-order differential equations are both statistically and biologically preferable to first-order kinetics (Deans *et al.*, 1986, on agricultural soils; Marion and Black, 1987, on tundra soils). Talpaz *et al.* (1981) pointed out that the estimation of N_0 is sensitive to the statistical techniques used to fit the models to the data.

Several studies with forest soils have determined N_0 and used it to compare the effects of harvesting. Burger and Pritchett (1984) found that N mineralized decreased with increasing severity of treatments in a slash pine (*Pinus elliottii* ENGELMN.) plantation: control, 5.5 mg N/kg soil; harvested and burned, 4.5 mg N/kg soil; harvested, burned, and plowed, 3.3 mg N/kg soil. These values represented from 4 to 6% of the total N in the soils. The laboratory incubations did not account for the importance of microclimatic differences among the treatments, so Burger and Pritchett used adjustments for temperature and moisture effects on N mineralization to estimate likely mineralization under ambient conditions. These adjustments reversed the order of the treatments, with the harvested, burned, and plowed treatment showing the highest estimated N mineralization. The value of even very intensive incubations under laboratory conditions is clearly limited where microenvironmental differences among sites are important.

Fox *et al.* (1986) performed a similar experiment on the effects of harvesting and site preparation on four loblolly pine (*Pinus taeda* L.) watersheds. The trend among treatments was less clear than in the slash pine study. Mineralization in the harvest + disking treatment exceeded the rate in the control watershed, which exceeded the rate in the two other site preparation treatments. These authors also used on-site incubations in buried bags to examine microclimate effects, and they also found that rates in the control stand were lower than in any harvest and site preparation treatments.

Maimone *et al.* (1987) determined N_0 for fertilized and unfertilized plots in a 13-year-old plantation of loblolly pine. One year after N fertilization, N_0 in surface soil samples was greater in fertilized plots. Fertilization with P did not alter N mineralization. They found substantial seasonal variability in N_0, which underscores the fact that N_0 is not a robust measure of any truly "mineralizable" portion of soil N. Similar findings have been reported for agricultural soils (see Bonde and Rosswall, 1987).

Fyles and McGill (1987) examined N mineralization in eight stands of jack pine (*Pinus banksiana* LAMB.) and of white spruce [*Picea glauca* (Moench) Voss] in Alberta during 37 weeks of incubation. Samples from A horizons mineralized from 6 to 16% of the total N present, and N_0 ranged from 12 to 43% of the total N. Nitrification began as early as Day 7 of the incubation for one site, but demonstrated lag times of several months for some others. The percentage of the net mineralized N that was also nitrified ranged from <1 to > 80%.

A similar procedure has been developed to examine the rate of nitrification during prolonged incubations, but no presumption is made about the shape of the curves that result. In this "microlysimeter" (Melillo 1977) method, samples are repeatedly leached with water to remove water-soluble NO_3^-. Matson and Vitousek (1981) used the microlysimeter method to examine the time course of NO_3^- production in control and clear-cut hardwood forests. They found that clear-cut forests generally showed higher rates of nitrification and less lag time before the maximum rate of NO_3^- production was reached. Thorne and Hamburg (1985) used this method to examine nitrification in an old-field successional sequence in New Hampshire. They found that nitrification decreased with stand age and that the lag time before initiation of nitrification increased with stand age. Gosz and White (1986) found that nitrification was delayed in microlysimeter incubations of soils from four forest types in New Mexico, but after about 3 weeks, soils from aspen and mixed conifer stands showed significant nitrification.

K. Nadelhoffer (personal communication) has improved on the basic method by designing a system that allows the evolution of CO_2 to be monitored during the incubations.

4. Microbial Biomass N

The pool of N contained in microbial biomass is a major source of labile N in most soils. Despite the fact that several N availability estimates are affected by the microbial-N pool, little attention has been placed on using the microbial-N as a separate entity for N fertility assessments. Recent experiments using ^{15}N have shown that for many soils the microbial biomass constitutes a major source of available N (Paul and Juma, 1981; Myrold and Tiedje, 1986). Marumoto and coworkers (1982a, 1982b) and Amato and Ladd (1980) also have shown that N turnover from dead microbial cells is about five times faster than from nonbiomass-N in the soil, which underscores the potential importance of the microbial pool as a readily available source of N.

There are many methods for estimating the size of the soil microbial biomass but fewer for the estimation of biomass-N. The two most common methods are the chloroform fumigation incubation method (CFIM) (Jenkinson and Powlson, 1976; Voroney and Paul, 1984) and chloroform fumigation followed by immediate soil extraction (the "direct extraction" method of Brookes *et al.*, 1985). In both methods, chloroform fumigation is used to lyse microbial cells and release protoplasmic constituents within the soil, but the technique employed to recover the released microbial-N differs.

In CFIM, biomass-N is estimated from the flush of N (F_N), that is mineralized during a 10-day incubation at 20 to 25 C, divided by a factor (K_N) that accounts for microbial N not mineralized (or reimmobilized) during the postchloroform incubation. In some cases, the flush of N may be reduced by the amount of N mineralized in a paired control (nonfumigated) incubation. The K_N factor is either assumed to be constant (Shen *et al.*, 1984, where K_N was set equal to 0.68) or to be a function of the ratio of net C mineralized to net N mineralized (Voroney and Paul, 1984). This ratio is used to account for differences in the amount of microbial N immobilized during the incubation period.

Direct extraction of labile N after fumigation is simpler than the 10-day incubation. The extraction uses K_2SO_4 to extract N from lysed cells immediately after chloroform fumigation. At the time of fumigation, a separate soil subsample is also extracted with K_2SO_4. The K_2SO_4 extracts are then digested and analyzed for NH_4^+-N. The difference between the total N contents of the fumigated and unfumigated subsamples is then adjusted to a biomass-N values using a K_N value (Brookes *et al.*, 1985).

More evidence of the significance of the size of the microbial N pool on N availability has been demonstrated by relating biomass-N values with other N availability estimates and plant uptake of N. Myrold (1987) showed a strong correlation between biomass-N and net N mineralized during anaerobic incubation for several forest soils ($r^2 = 0.7$). He also labeled microbial biomass with ^{15}N and found that anaerobically mineralized-N and biomass-N pools had similar atom percentage enrichments of ^{15}N [where atom percentage enrichment of ^{15}N = (^{15}N atoms/(^{14}N atoms + ^{15}N atoms)) × 100%]. This result further suggests that the anaerobic incubation method provides an estimate of the size of the microbial biomass, thus the amount of N contained in soil microbes may be useful in estimating N availability.

Carter and Rennie (1982), Carter (1986), and Carter and Macleod (1987) also found that biomass-N correlated well with N_0 values for several agricultural and grassland soils ($r^2 = 0.8$ to 0.9). Further, Azam and others (1986) found that biomass-N correlated well with N uptake by wheat plants ($r^2 = 0.9$), again suggesting that a major portion of plant available N is derived from mineralization of dead microbial biomass (Jenkinson and Ladd, 1981).

Ideally, estimates of microbial biomass-N should only be sensitive to the size of the microbial-N pool. However, both of these techniques include a variable portion of the nonmicrobial labile-N pool (Figure 1). Because they are laboratory methods, they are also insensitive to fluctuations in field soil moisture and temperature. Studies are needed to assess seasonal variation in the size of forest soil microbial-N pools and how they relate to seasonal variation in in-field N mineralization. These estimates of microbial biomass-N also reflect only the pool size of microbial-N and not the flux through the pool, and the flux is probably more important in regulating N availability in the field. Indeed, the measures do not distinguish between major taxonomic groups (such as fungi and bacteria) that may differ strongly in the relationship between microbial N and the flux of N through the pool. Estimates of microbial N using chloroform methods may afford no great advantage relative to other methods of assessing N availability for applied assessments of forest nutrition. Estimates of microbial-N usually correlate well with simpler methods, and the chloroform techniques are more complicated and potentially hazardous (chloroform is a carcinogen). However, the application of ^{15}N techniques to microbial-N methods can greatly enhance the

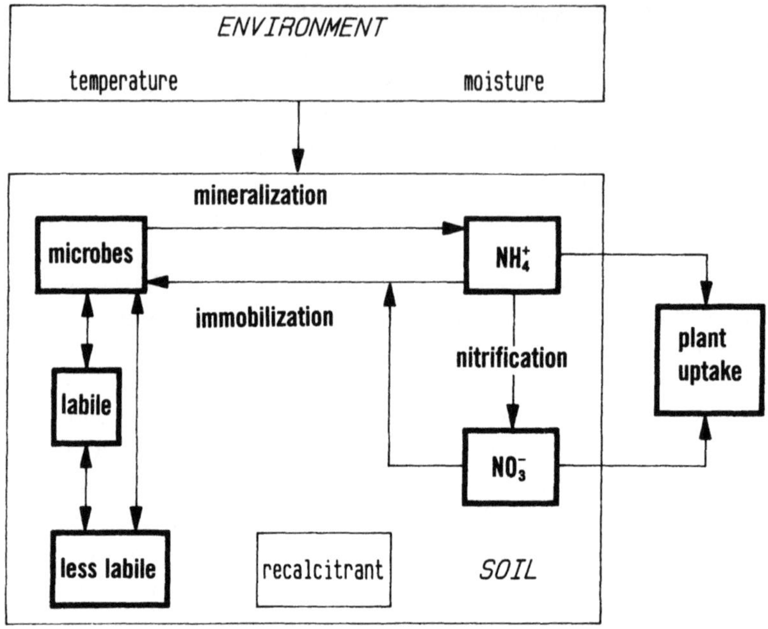

Figure 9. Bioassays integrate all soil components except the recalcitrant component, temperature and moisture are usually held constant, and soil preparation may alter mineralization/immobilization patterns. Ecosystem N budgets are similar, but they include on-site environmental conditions.

basic understanding of the limitations of each component on N availability; these approaches are discussed later.

D. Bioassays and Ecosystem N Budgets

Plants cultivated in soil samples provide an integrated measure of the N mineralized during the growing period, as does N uptake by vegetation in intact ecosystems (Figure 9). No standard protocol is available for either bioassays or ecosystem N budgets. Critical factors in bioassays include choice of bioassay species (see Kadeba and Boyle, 1978), length of bioassay period, pretreatment of soils (especially sieving), and the environmental conditions of the bioassay. Components of ecosystem N budgets differ in importance among ecosystems, but the most important items are determining the N content of litterfall, of fine root turnover, and of the increment of N in biomass.

Lamb (1975) used a bioassay with radiata pine (*Pinus radiata* D. Don) to examine mineralization from forest floor materials collected in 12 plantations of radiata pine. He mixed the forest floor material with perlite (5:4 by volume) and planted four seeds in each pot, with eight replicate pots for each sample. After 25 weeks of growth, N uptake ranged from 43.5 mg-N/pot to 77.5 mg-N/pot. Uptake of N did not correlate significantly with the total N content of the forest floor materials, but it correlated very well ($r^2 = 0.8$) with the N mineralized aerobically in the laboratory over 90 days.

Youngberg (1978) examined the ability of forest floor materials from nine types of Douglas-fir sites to supply N to two croppings of Douglas-fir seedlings. He found that litter quality had greater influence on the mineralization of N than did the total amount of N added.

Safford (1982) examined the ability of a bioassay with paper birch (*Betula papyrifera* Marsh.) to predict the response of birch stands fertilized in the field. He found that nutrient uptake in the bioassay generally correlated well with growth response to fertilization, with r^2 in the range of 0.4 to 0.8. Mead and Pritchett (1971) performed a similar experiment with slash pine and found that N uptake by seedlings correlated moderately (r^2 of 0.3 to 0.5) with stem growth in the field. Both these studies used the bioassay as an indicator of nutrient limitation of growth rather than as indexes of nutrient supply rate.

The most accurate method for estimating the N supply in an ecosystem is measuring the major components of the annual N cycle: inputs, uptake by plants, and outputs. These processes integrate all the factors that regulate N availability. Aber and others (1985) calculated the N uptake in aboveground tissues in 13 forests and compared these rates with mineralization estimates derived from on-site incubations in buried bags. The buried-bag estimates were higher than the N budget estimates of available N, and Aber *et al.* used this difference to calculate what the production of fine roots must have been to account for the discrepancy in the two N availability methods. They felt that the N mineralization estimates were accurate; any bias in the estimates would greatly alter the estimate of fine root production. They concluded that although the average fine root biomass at any single time was lower on sites with high N availability, the annual production of roots on such sites was very high owing to rapid turnover of fine roots. In all other studies of root production and N availability, root allocation has been estimated to decrease as N availability increases (see Vogt *et al.*, 1986). If the pattern described by Aber and others were real, it would be a major finding in ecosystem studies that was based on simple assays of N availability.

A complete tally of ecosystem N fluxes is expensive, and it is possible to obtain an index of the ecosystem N budget by measuring only a part of the whole cycle. Sites with greater N availability generally have more N cycling in litterfall. For example, Hedman and Binkley (1988) found that the N content of litterfall in 11 hardwood forests in the Piedmont of North Carolina ranged from 35 to 68 kg-N/ha annually. Vitousek (1982) found a clear relationship between the N in litterfall and litter quality (indexed as the C/N of the litter) in a review of the literature and concluded that a positive feedback may exist between N availability, litter quality, and future N availability.

Pastor and Post (1986) developed a simulation model that explicitly incorporates this feedback between litter quality and N availability. The model predicts changes in soil organic matter (SOM), forest productivity, and species composition during succession across gradients in soil moisture in a wide variety of conifer and hardwood forests in eastern North America. Further simulations

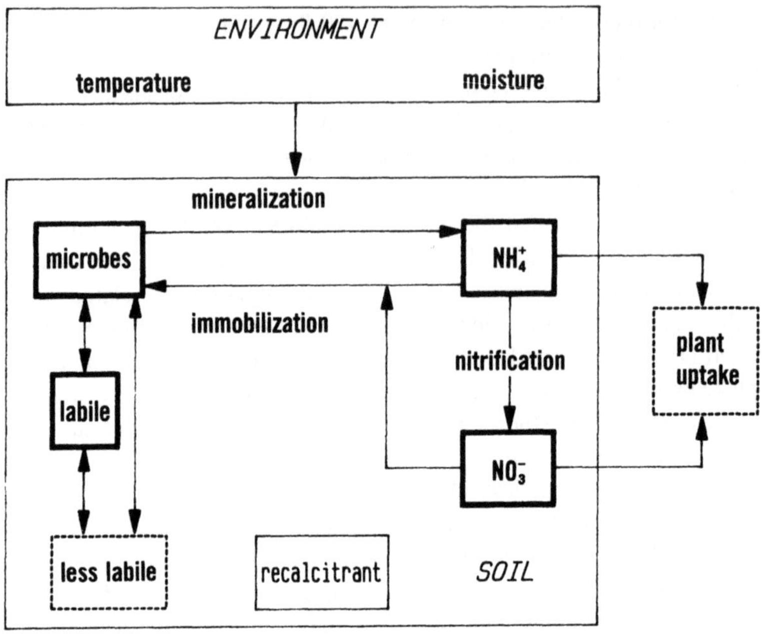

Figure 10. In-field methods include varying contributions from the inorganic, microbial, and labile components under the influence of on-site microenvironments. Resin bags are also sensitive to plant uptake, and longer-term incubations may include contributions from the less-labile component.

have shown that predictions of dynamics in boreal forests are very sensitive to this feedback (Pastor *et al.*, 1987b). Experimental evaluation of this positive feedback (and other biological and ecological constraints) will require clearer understanding of the components responsible for the estimate of N availability derived from each assay method.

E. Field Incubations

As noted earlier, nitrogen mineralization is strongly affected by microenvironmental conditions (Figure 10). A variety of methods have been developed to account for the importance of environmental factors.

1. Ion Exchange Resin Bags

Ion exchange resins have been used in laboratory assays of P availability for several decades (see Amer *et al.*, 1955; Sibbeson, 1977), and they have recently been used for evaluation of nutrient availability under field conditions in forests (Binkley and Matson, 1983), grasslands (Gibson *et al.*, 1985), and deserts (Lajtha, 1988). These field assays involve placing a known weight or volume of resins into net bags, incubating the bags in the soil for 1 month to a year, and extracting the resins with a salt solution to determine the quantity of ions adsorbed. Laboratory studies have shown that the accumulations of ions on the resin bags depend strongly on nitrogen mineralization rate, ion form (NO_3^- moves to the bags with less resistance than does NH_4^+), water movement to the bags, and competition for nutrients with microbes and plants (Binkley, 1984b). Single extractions of resins may not remove all adsorbed N; repeated extractions or determination of a recovery factor may be necessary if complete recovery is needed.

In forest soils, Binkley and Matson (1983) used resin bags to evaluate the effects of N-fixing Sitka alder [*Alnus sinuata* (Regel) RYDB.] and red alder (*Alnus rubra* BONG.) on N availability in a Douglas-fir plantation. Under greenhouse conditions, they found that resin bags in soils from the alder portions of the plantations had about twice as much N as those from the no-alder portions. Resin bags placed in the field for 11 months showed the N availability in the Sitka alder portion was ten times greater than in the no-alder portion, and the red alder portion had twice the availability of the Sitka alder portion. Only a trace of NO_3^- accumulated on the resins in the no-alder portion of the stand. About 20% of the inorganic N on the Sitka alder resins was NO_3^- compared with 50% on the red alder resins.

Binkley (1984a) examined the effects of clear-cutting on N availability at three elevations on Vancouver Island, British Columbia. Anaerobic incubations showed no change in N availability after cutting, but ion exchange resin bags showed that cutting increased N availability by sevenfold to twentyfold. Cutting apparently had no effect on the chemical quality (mineralizability) of the soil, but either the microclimatic effects favored mineralization or reduced uptake after cutting left more NH_4^+ and NO_3^- in the soil to accumulate in the resin bags. Resin-bag estimates of N availability depend on transport of ions to the bags (Binkley, 1984b), and increased soil moisture in the clear-but sites may have increased ion transport to the bags. The combined importances of mineralization, immobilization, plant uptake, and ion transport make it difficult to interpret the mechanisms contributing to resin estimates of N availability. These factors also affect the supply of N to root surfaces, although the relative importances may differ.

Hart and Binkley (1985) compared N availability in unfertilized plots of loblolly pine with adjacent plots that had been fertilized 2 years previously with N + P. They found that mean values for fertilized plots exceeded those of the controls in 9 of the 11 trials. Other methods of assessing N availability (aerobic and anaerobic incubations; boiling salt extraction) showed higher availability in 4 to 8 of the 11 trials. The resin bags showed the NO_3^- comprised about one-third of the N supplied in both fertilized and unfertilized plots, whereas aerobic incubations produced net negative values for nitrification. Hart *et al.* (1986) also tested the ability of resin bags to predict the response of loblolly pine stands to N + P fertilization. They found no significant relationship between N or P accumulated on the resins and the growth response of stems; no other measure of N availability proved successful either.

Binkley *et al.* (1986) placed resin bags in a series of 13 ecosystems where N budgets and mineralization in buried bags had been assessed. After a single growing season, the N accumulated on the resin bags correlated well with the other measure. Two sites showed anomalously high NO_3^- on the resin bags; these sites were the only ones that had shown substantial leaching losses of NO_3^- in the earlier studies. The most interesting result of this study was that resin bags accumulated a significant amount of NO_3^- on sites where incubations had shown none (owing to either nitrification or nitrate in throughfall) and that resin bags accumulated significant amounts of NH_4^+ on sites where incubations showed 100% nitrification of mineralized N. These findings again emphasize the need for research to identify which components in Figure 1 contribute the estimates of available N for each method.

Krause and Ramlal (1987) compared nutrient availability in an 80-year-old forest of balsam fir (*Abies balsamea* (L.) MILL.) and black spruce [*Picea mariana* (Mill.) B.S.P.] with an adjacent 2-year-old clear-cut site. Resin bags in the clear-cut site adsorbed several times more NH_4^+ and NO_3^- (and phosphate) than bags in the older forest.

R.H. Waring *et al.* (1987) used resin bags to examine changes in the availability of N across a mortality sequence of mountain hemlock [*Tsuga mertensiana* (Bong.) CARR.] ecosystem. Inorganic N accumulated on the resin in the bare and young-regrowth zones was about four times greater than in the old-growth portion of the sequence; the older-regrowth portion fell in between.

Resin bags have appeared to be very successful in showing differences in N availability among ecosystems and among treatments. The major weakness of the method is that the components responsible for observed differences (such as differences in mineralization, uptake by plants and microbes, and ion mobility) cannot be identified.

2. Buried-Bag Incubations

Incubations of soils in buried plastic bags has become the most widely used method of assessing N availability under field conditions. This method was developed in the later 1950s (Eno, 1960) and used extensively in Europe in the 1960s and 1970s (Antonietti, 1968; Runge, 1971, 1974; Ellenberg, 1977; Glavac and Koenies, 1978a, 1978b; Kovacs, 1978). Forest soil scientists in North America have never adopted this method (it was not mentioned in Keeney's 1980 review), but ecosystem ecologists started using the method in the late 1970s (Matson and Vitousek, 1981; Melillo, 1981). Soil samples are typically taken with a soil corer and placed in plastic bags with as little disturbance as possible. The bags (15- to 30-μm thick) allow transfer of oxygen and carbon dioxide, but prevent loss of liquid water (Gordon *et al.*, 1987). The bags are buried beneath the forest floor and incubated for 1 to 2 months (or over the winter period). After the incubation period, NH_4^+ and NO_3^- are determined in 2 *M* KCl extracts. Net mineralization is calculated as the sum of NH_4^+ and NO_3^- in the extract, minus the amounts present before incubation. Annual patterns of mineralization are obtained by repeating the incubations. The method is superior to laboratory incubations, primarily in its sensitivity to on-site temperature regimes. It may also be more sensitive to on-site moisture levels as long as the soil moisture content at the beginning of each incubation period is representative of the conditions for the whole incubation period.

In some studies, subsamples of large, composited samples are used to answer specific questions. For example, Matson and Vitousek (1981) used composited samples for reciprocal transplants into cut and uncut sites in Indiana. They found that transplanting soil from a control forest into a recent clear-cut forest increased net mineralization by 65%. Transplants into older clear-cuts showed no effect, clearly indicating the importance of elevated temperature on mineralization in the clear-cut.

Runge (1971) characterized the mineralization patterns with soil depth in a 125-year-old beech (*Fagus sylvatica* L.) forest in Germany. Of the total mineralization of 112 kg-N/ha annually, about two-thirds occurred in the forest floor (Figure 2). Ellenberg (1977) summarized mineralization data for a variety of European forests (Table 2). The range in rates within a single forest type was large relative to differences among forest types. Glavac and Koenies (1978a, 1978b) were the first to use buried bags to examine the effects of clear-cutting on mineralization. They found that mineralization tripled after cutting of two spruce stands, but the rate was unaffected by clear-cutting of a beech stand. Virzo De Santo *et al.* (1982) found that N mineralization in a Mediterranean beech forest was about 43 kg-N/ha annually compared with a rate of 47 kg-N/ha for a beech forest mixed with fir (*Abies alba* Mill.). Gordon and Van Cleve (1983) used buried-bag incubations to examine the effects of harvesting on mineralization in white spruce [*Picea glauca* (Moench) Voss] forests. From 12 to 32% of the bags showed a net decrease in NH_4^+ during the incubations. Clear-cutting increased net mineralization, and greatly increased nitrification. Mladenoff (1987) estimated nitrogen mineralization in old-growth maple (*Acer saccharum* L.) and hemlock [*Tsuga canadensis* (L.) CARR.] forests in Michigan, and contrasted the rates with those in treefall gaps. Gaps in the maple part of the forest showed reduced mineralization, whereas gaps in the hemlock portions showed increases in both net mineralization and nitrification.

Nadelhoffer *et al.* (1983) examined patterns in N mineralization and N uptake in nine forests in the Arboretum of the University of Wisconsin. They found that mineralization rates from buried-bag incubations did not correlate with total soil N, C, or C/N, but the rates correlated very well with N uptake by the trees ($r^2 = 0.9$) and N content of litterfall ($r^2 = 0.6$). Pastor *et al.* (1986) performed a similar study on undisturbed forests at Blackhawk Island, Wisconsin, where differences in geomorphology provided a range of soil and vegetation types on this small island. Annual rates of mineralization in the old (> 100 years), undisturbed forests ranged from 26 kg-N/ha in a red pine (*Pinus resinosa* AIT.) stand to 84 kg-N/ha in a sugar maple (*Acer saccharum* MARSH.) stand. Pastor and others calculated net nitrification as the NO_3^- present in buried bags at the end of an incubation sequence, minus the NO_3^- present at the beginning. On a monthly basis, the rate of nitrification may exceed net mineralization if all mineralized N is nitrified and if there is less exchangeable NH_4^+ after incubation than before. In fact,

Table 2. Annual mineralization estimates from in-field incubations (kg/ha)

Method/forest type	Conditions	Horizon or depth (cm)	Ammonium	Nitrate	Sum	Percentage of total N	Reference
Buried-bag incubation							
Beech	125-yr-old	0 + 0–14	47	65	112		Runge 1974
Norway spruce	47-yr-old uncut	0 + Ah			72		Glavac and Koenies 1978a
	clear-cut				200		
Norway spruce	88-yr-old uncut	0 + Ah			31		Glavac and Koenies 1978b
	clear-cut				94		
Beech	88-yr-old uncut				35		Glavac and Koenies 1978b
	clear-cut				35		
Beech	Various				24–238		Sources in Ellenberg 1977
Maple/ash	Various				100–173		
Ash/maple	Various				147–377		
Elm	Various				29–106		
Oak/ash	Various				24–115		
Oak/birch	Various				40–81		
Sitka spruce	35-yr-old	0					Harmer and Alexander 1985
	untrenched				12		
	trenched				17		
Beech	95-yr-old	0–7			43	0.9	Virzo De Santo *et al.* 1982
Beech/fir	75-yr-old				47	1.0	
Black oak	>35-yr-old	0–10			111	7.9	Nadelhoffer *et al.* 1983
White/black oak					91	4.8	
Red oak					100	2.8	
Sugar maple					62	3.5	
Paper birch					51	2.5	
White pine					80	4.9	
Red/white pine					50	3.4	

White spruce					47	2.7	
Red pine					32	1.8	
Red pine	> 100-yr-old	0–15	4	22	26		Pastor *et al.* 1984
White pine			20	19	39		
Red oak			34	19	53		
Red oak			−15	75	60		
White oak			65	2	67		
Sugar maple			3	75	78		
Sugar maple			−14	98	84		
Hemlock			26	3	29		
Oak savanna	> 100-yr-old	0–10			65	2.9	Pastor *et al.* 1978a
Maple	Old growth	0–10	34	68	102		Mladenoff 1987
	Gap		30	59	89		
Hemlock	Old growth		62	27	89		
	Gap		37	70	107		
Red spruce-balsam fir	Young	?	42	1	43		Thorne *et al.* 1987
	Mature		53	1	54		
	Old		17	1	18		
Mixed conifers	Regrowth	0–7.5	−5	42	37	1.7	Hart and Firestone (1989)
	Old growth		11	5	16	0.6	
Loblolly pine	22-yr-old	0–15	21	4	25		Vitousek and Matson 1985
	Stem harvest		3	97	100		
	Whole tree		13	82	95		
Loblolly pine	< 20-yr-old		18	0	18		Satterson 1985
	40–60-yr-old		43	1	44		
	80–100-yr-old		29	0	29		
Oak-hickory	> 150-yr-old		75	0	75		

Table 2. *Continued*

Method/forest type	Conditions	Horizon or depth (cm)	Ammonium	Nitrate	Sum	Percentage of total N	Reference
Pitch pine	Upland mesic	0–15	36	2	38	Poovarodom *et al.* 1988	
			50	3	53		
Mountain hemlock	Old growth	O2 + 0–15			2.5	0.2	Matson and Boone 1984
	Dead				5.4	0.5	
	Young regrowth				4.5	0.4	
	Old regrowth				2.6	0.2	
Lodgepole pine	80–110-yr-old	0–15	7.6	0.4	8	0.5	Fahey *et al.* 1985
Tropical hardwoods	Old growth	0 + Ah	5–150	23–590	128–634	3–24	Pfadenhauer 1979
Tropical hardwoods	75-yr-old	0–15			760		Matson *et al.* 1987
	1-yr regrowth				1140		
	6-yr regrowth				588		
Closed-tube incubation							
Norway spruce	74-yr-old	F horizon			48	3.5	Van Praag and Weissen 1973
Radiata pine	10-yr-old						
	+ fertilizati-	0–40			16		Raison *et al.* (1987b)
	on				43		
Mixed conifer	100-yr-old	0–14			12		Frazer *et al.* (in review)
	6-yr-old				45		
	18-yr-old				28		
Resin-core incubation							
Mixed oak	Low elevation	0 + 0–14	24	24	48		Powers 1987
Mixed conifer	Midelevation		19	40	59		
Red fir	High elevation		15	5	20		

Beech	100+-yr-old	0–15	10–20	60–70	80–90		Bloss and Binkley (in review)
Fraser fir	Mature	0–15	65	33	98		Sasser and Binkley 1988
	Dead zone		47	26	72		
	Young regrowth		30	6	36		
	Older regrowth		27	4	31		
Balsam fir	Mature	0–15	82	8	90		Sasser and Binkley 1988
	Dead zone		90	32	112		
	Young regrowth		80	1	81		
	Older regrowth		69	3	72		
Fraser fir/red spruce	>50 yr	0–15			28–180		Strader *et al.* 1988
White spruce	Old	0–7.5	3	−0.5	2.5	0.8	Hart and Gunther (in review)
Thinleaf alder	Old		13	9	22	1.4	
Mixed conifer	Regrowth	0–7.5	5	29	34	1.6	Hart and Firestone 1989
	Old growth		14	6	20	0.7	
Douglas-fir	55 yr, infertile	0–15	4.4	−3.9	0.5		Binkley *et al.* (in review)
Red alder/Douglas fir			27	99	126		
Douglas-fir	55 yr, fertile		15	14	29		
Red alder/Douglas-fir			9	115	124		

two of their eight sites showed more net nitrification on an annual basis than net mineralization, indicating that the soils had less exchangeable NH_4^+ at the end of the 1-year period. On these eight sites, annual net mineralization did not correlate significantly with soil total N, but correlated strongly with the contents of N ($r^2 = 0.7$) and P ($r^2 = 0.8$) in litterfall. In contrast, Pastor *et al.* (1987a) found positive correlations between total soil N and estimates from incubations in buried bags for abandoned agricultural fields being invaded by forests.

Matson and Boone (1984) used buried bags to assess N availability in the same mortality sequence of mountain hemlock examined by R.H. Waring *et al.* (1987). They found that N availability in the bare zone was double that of the old-growth forest and the rate was intermediate in a young-regrowing forest. These results showed the same pattern as the resin bags (R.H. Waring *et al.*, 1987), although the relative magnitudes of the differences across the sequence varied.

The effects of root exclusion (through soil trenching) on N mineralization in a Sitka spruce [*Picea sitchensis* (Bong.) CARR.] forest were examined by Harmer and Alexander (1985). They reported that mineralization in trenched plots was 50% greater than in untrenched plots. This highlights the importance of the turnover of fine roots in the balance between mineralization and immobilization.

Vitousek and Matson (1985) found that harvesting and various site-preparation treatments increased buried-bag mineralization estimates by fivefold in a loblolly pine study. Nitrification was increased from about 10% of total net mineralization to essentially 100%. They also compared these values with resin bags for a 1-month period. Net mineralization in the bags correlated poorly with the $NH_4^+ + NO_3^-$ on the resins, but net nitrification correlated moderately well with NO_3^- on the resins ($r^2 = 0.5$). Annual sums based on buried-bag incubations and laboratory incubations were well correlated (r^2 of about 0.8 for net mineralization and for net nitrification).

Satterson (1985) examined mineralization and nitrification in a replicated sequence of old-field succession in the Piedmont of North Carolina. The rate of net mineralization showed no clear trend with successional stages, although the mineralization of the forest floor became proportionately more important with age. As in the studies by Nadelhoffer *et al.* (1983) and Pastor *et al.* (1984), Satterson found that net mineralization correlated very highly with the N content of litterfall ($r^2 = 0.7$).

Gosz and White (1986) examined N mineralization along an elevational gradient in New Mexico. They found large differences between seasons and between ecosystems, but they did not provide a complete annual pattern. Montagnini *et al.* (1986) examined in-field nitrification for on month and found that rates in a young stand dominated by N-fixing black locust (*Robinia pseudoacacia* L.) were four times higher than in adjacent mixed hardwood stands. The black locust stand also showed high NO_3^- concentrations (from 1 to 10 mg N/L) in soil solution, compared with trace concentrations (< 0.1 mg N/L) in the hardwood stands.

Pfadenhauer (1979) measured N availability in six tropical hardwood forests and found that both the rate and the proportion of the mineralized-N nitrified differed markedly among forests. His highest rate was over 600 kg-N/ha in the forest floor plus Ah horizons, a phenomenal rate compared to temperate forests. Matson *et al.* (1987) used the buried-bag method to evaluate N availability in some tropical forests. Clear-cutting and burning a relatively fertile forest in Costa Rica led to a 40% increase in N mineralization (Matson *et al.*, 1987). The mineralization rate in the intact, 75-year-old forest was on the order of 750 kg-N/ha annually.

3. Closed-Top Tube Incubations

In the 1950s, Lemée (1967) developed a technique for incubating soil samples in aluminum cans that were perforated on the sides, closed on top, and open on bottom. This design keeps soil moisture in the cans similar to that of the bulk soil, prevents direct uptake by roots, and prevents direct leaching of the soil by rainfall. He found that mixing soils before incubation increased the mineralization rate by 50% and that in-field incubations mineralized only about 4 to 50% of the N mineralized in laboratory incubations. Rapp *et al.* (1979) used Lemée's technique to estimate N mineralization in an Italian stone pine (*Pinus pinea* L.) forest near the Mediterranean Sea. The forest floor plus A1 horizon mineralized about 11.5 kg-N/ha annually, roughly matched by the N transferred to the forest floor in throughfall. These two fluxes combined to equal only about half of the estimated N uptake by the forest, indicating that perhaps 23 kg-N/ha were mineralized annually in deeper soil horizons.

Adams and Attiwill (1986a, 1986b) modified this method by using plastic polyvinyl chloride (PVC) tubes capped with inverted petri dishes. They found that the caps prevented a small but significant amount of NO_3^- loss during 2-month incubations. In-field rates of mineralization were lower than laboratory rates owing to suboptimal temperature and moisture conditions in the field. This method indicated that N mineralization was doubled following bushfires in *Eucalyptus* forests.

Raison *et al.* (1987a) performed a variety of experiments to determine the importance of factors affecting mineralization in closed-top cores. They found that sieving soils before incubation increased mineralization rates by twofold to tenfold, including a more-than-proportional increase in nitrification. Soils that had been fertilized earlier with 200 kg-N/ha showed fourfold greater mineralization in closed-top tubes than unfertilized soils. In contrast, laboratory incubations on sieved soils showed no effect of fertilization. They also found that metal tubes did not alter soil temperatures and gave the same mineralization rates as PVC tubes. The rate of mineralization was relatively insensitive to the length of the incubation period, indicating no major disturbance effect of inserting the cores. The cores were inserted to a depth of 40 cm, and Raison *et al.* determined that about one-fourth of the mineralization in two forests occurred between 15 and 40 cm (Figure 2). This depth interval could not be sampled practically with the buried-bag method.

Raison *et al.* (1987b) followed up on their earlier work by using closed-top cores to evaluate the effects of fertilization on a 10-year-old plantation of radiata pine (*Pinus radiata* D. Don.). One year after fertilization, mineralization in the control stand was only 16 kg-N/ha annually compared with 43 kg-N/ha in the fertilized plot. Three years after fertilization, mineralization in the control plots declined to 7 kg-N/ha annually, whereas the fertilized plots remained at 24 kg-N/ha.

Frazer *et al.* (in review) used closed-top cores to examine changes in mineralization following harvesting. They used a 100-year-old mixed conifer forest, an adjacent area clear-cut 6 years previously and planted to ponderosa pine (*Pinus ponderosa* Laws.), and an adjacent stand of ponderosa pine that regenerated naturally after a clear-cut 18-years previously. Mineralization decreased with stand age and inhibition of nitrification appeared to occur in the forest floor of all stands.

The closed-top core method is simpler than any in-field method except resin bags, and the prevention of uptake by plants is a potential advantage over resin bags. Additional research comparing the contribution of the various components in Figure 10 to the mineralization estimate might provide valuable insights for assessing the relative accuracy of in-field methods.

4. Resin Cores

Di Stefano (1984) and Di Stefano and Gholz (1986) combined the resin bag and core methods in a system that incorporates in-field incubations in open tubes with bags of ion exchange resin on the top and bottom. The resin bag on top of the tube deionizes throughfall (or leachate from overlying soil horizons) entering the tube, and the resin bag on bottom captures any ions leaching from the core. After the incubation period, the resin bag on top is discarded (unless a measure of input to the core is desired), and the soil and bottom resin bag are extracted with KCl. The extractable NH_4^+ and NO_3^- from the soil and the bottom resin bag estimate N mineralized during the incubation. This technique is potentially superior to the buried-bag method because it allows moisture in the core to fluctuate through the incubation period. It may also be better than both the buried-bag and closed-top core methods by allowing the products of mineralization to be removed from the core. This would minimize any feedback effect of NH_4^+ and NO_3^- accumulation on mineralization and immobilization processes.

Di Stefano (1984) found that clear-cutting a 29-year-old slash pine plantation increased the monthly mineralization rate from near 0 to about 3 mg-N/kg soil, which roughly matched the rate found in a 9-year-old plantation. He also found that positive rates of net mineralization in the summer were partially offset by negative rates in spring and fall incubations.

Powers (1987) used this method to examine patterns of N mineralization with elevation in California. Mineralization was highest in the midelevation mixed conifer ecosystem, followed by the warm, low-elevation mixed oak ecosystem. The lowest rate occurred in the cool, high-elevation red fir (*Abies magnifica* A. Murr.) ecosystem. In the midelevation site, the forest floor contributed about one-third of the net mineralization. The forest floor accounted for most of the mineralization at the other two sites. Most of the N mineralized in all sites in the mineral soil was nitrified, but most in the forest floor was not.

Strader *et al.* (1988) adopted the resin core method for evaluating mineralization in high-elevation stands of red spruce (*Picea rubens* SARG.) and Fraser fir [*Abies fraseri* (Pursh) POIR.] in the Southern Appalachian Mountains. They used four sequential periods of incubations to provide an annual estimate of mineralization in 19 forests. The rates ranged from 14 to 25 mg-N/core on an annual basis, which extrapolated to a range of 26 to 180 kg-N/ha annually when the variable rock contents of the soils were included. NO_3^- represented from 1 to 60% of the N mineralized. Most of the mineralized N (from 50 to 85%) was found on the resins, indicating the potential importance of allowing the products of mineralization to be removed from the cores. In a related study, Sasser and Binkley (1988) examined the patterns in mineralization across a mortality sequence (fir wave) of Fraser fir and another of balsam fir [*Abies balsamea* (L.) MILL.]. In both sequences, mineralization was highest in the dead zone, followed by lower rates in regenerating zones, with a gradual increase as stands aged. In a third study, Bloss and Binkley (in review) compared mineralization in plots in a beech (*Fagus grandifolia* EHRH.) forest where rooting by feral boars (*Sus scrofa*) had been excluded. They found no differences between rooted and unrooted areas. Interestingly, their net mineralization rates were similar to those found in the related studies in spruce-fir forests, but nitrification was more important (80% of mineralized N) in the beech forests.

Binkley *et al.* (in review) measured the effect of 50 years of N fixation by red alder (*Alnus rubra* BONG.) in two conifer forests with sequences of resin core incubations every 2 months. At a very fertile site, resin cores (without resin bags on top) showed that mineralization (0- to 15-cm depth) in the alder/conifer stand was about 125 kg-N/ha annually compared with only 30 kg-N/ha annually in the conifer stand. Nitrate comprised about 95% of mineralized N in the alder/conifer plot compared with about 50% in the conifer plot. At a less fertile site, the mineralization rate in the conifer/alder stand was still about 125 kg-N/ha, with 80% of the mineralized N as NO_3^-. The conifer stand at the infertile site showed less than 1 kg-N/ha net mineralization; many months showed net immobilization. These authors also compared mineralization in resin cores with rates for adjacent cores placed in plastic bags. Net mineralization rates were similar for the methods but the resin cores showed greater nitrification. Interestingly, immobilization of added $^{15}NH_4$ during the 30-days incubation in the field was identical for the two methods, but the buried bags immobilized 40 to 95% more $^{15}N_3$ than the resin cores. Apparently, the removal of NO_3^- (by leaching to the resins) reduced the amount of NO_3^- immobilized during the incubation. This suggests that buried bags could underestimate the net nitrification that would occur in unconfined soil where tree uptake or leaching of NO_3^- might limit NO_3^- immobilization.

Hart and Firestone (1989) recently compared seasonal patterns of N availability from ion exchange resin bags, resin cores, and buried-bag incubations in young and mature stands of mixed conifers in California. The young forest showed about twice the rate of net mineralization in the resin cores and much greater nitrification. Buried-bag incubations provided very similar rates. The water content of the resin cores tended to be greater than in surrounding soil. Buried bags showed an opposite problem, that is, significant evaporation of water from within the bags during the incubation period with no opportunity for recharge from rainfall.

Clearly, much more work is needed on the effects of each method of assessing N availability on the components of mineralization and immobilization.

F. ^{15}N Methods

1. Natural Abundance

Nitrogen (N_2) in the atmosphere is 99.6337% ^{14}N and 0.3663% ^{15}N. The N in soils and ecosystems is less well mixed than the atmosphere and the percentage of ^{15}N commonly varies from 0.3653 to 0.3673%. Deviations from the atmospheric isotope ratios are usually measured as $\delta^{15}N$, which is the parts-per-thousand difference in the percentage of ^{15}N (Hauck and Bremner, 1976). If the $\delta^{15}N$ of two donor pools differ, then the relative contribution of each to a receptor pool can be identified. For example, Kohl *et al.* (1980) planted N-fixing soybeans (*Glycine max*) in a field where the $\delta^{15}N$ of available soil N (determined from non-N-fixing plants) was 7. The $\delta^{15}N$ of the soybeans was about 5 $\delta^{15}N$, indicating that about 30% of the soybean's N came from the atmosphere (with $\delta^{15}N$ of 0) and 70% came from the soil. When corncobs were added to the soil to increase microbial immobilization, the $\delta^{15}N$ of the soybeans dropped to 3, indicating that fixation supplied about 60% of the plant N.

The use of natural abundance of ^{15}N to trace N transformations in more complex systems is constrained by the amount of fractionation that may occur during transfer between pools. For example, Feigin *et al.* (1974) found that the $\delta^{15}N$ of grass crops was typically lower than the $\delta^{15}N$ measured for NO_3^- in laboratory incubations. Either the incubations did not adequately characterize the NO_3^- produced in the field or some fractionation of isotopes occurred in the uptake or transport of N within the grasses. In one of the few studies in forest ecosystems, Binkley *et al.* (1985) found that conifer foliage matched the $\delta^{15}N$ of soil NH_4^+ and NO_3^- at two sites but that foliage showed $\delta^{15}N$ much higher than available N at two other sites. Some researchers remain optimistic about the potential of using variations in the natural abundance of ^{15}N to trace ecosystem N dynamics (see B. Peterson and Fry, 1987), but more work is needed to establish the potential and limitation of this approach.

2. ^{15}N Fate

Additions of N compounds that are enriched or depleted in ^{15}N allow transfers among pools to be traced with much greater precision and certainty than is afforded by the natural abundance procedure (Hauck and Bremner, 1976).

Table 3. Summary of recent studies on the fate of ^{15}N-labeled fertilizers

Species	Age	N Applied ($kg \cdot ha^{-1}$)	Source	Duration (yrs)	Percentage recovered in Trees	Percentage recovered in Forest floor
Douglas-fir[1]	7–9	224	urea	2.0	25–36	21–31
Scots pine[2]	120–140	100	NH_4NO_3	2.0	12–29	13–22
Slash pine[3]	11	56	$(NH_4)_2SO_4$	1.5	25	9[a]
		224			27	6[a]
Corsican pine[4]	40	252	?	3.0	49	
		504			44	

[a] Includes only the litter layer.
Source: Data from
[1]Heilman *et al.*, 1982; [2]Melin *et al.*, 1983; [3]Mead and Pritchett, 1975; [4]Miller *et al.*, 1976.

a. Fate of Applied Fertilizer

Most use of ^{15}N in forest soil studies have focused on the efficiency of N-fertilizer applications or on the proportion of the applied-N that is taken up by crop trees. In 1975, Knowles reviewed this body of literature and noted that between 3 to 28% of the fertilizer-N was found in trees. Some of this variability can be accounted for by considering the different sources of N (urea and various NH_4^+ and NO_3^- salts), rates of application (50 to 224 kg-N/ha), and duration of these studies (2.5 to 24.0 months). A large portion of the applied-N was retained in the forest floor (14 to 56% for these same studies).

An overview of similar studies since Knowles's review substantiates these earlier findings (Table 3). Between 12 and 49% of the applied-N was recovered in the trees in these more recent studies, again with a significant portion of the N remaining in the forest floor (13 to 31%). Unfortunately, not all studies state clearly how the quantity of the tracer taken up was calculated. For example, Jansson and Persson (1982) pointed out that the uptake of fertilizer-N could be underestimated if the tracer-N mixed with native soil N prior to uptake by trees. This dilution of the isotopic enrichment of the fertilizer would be accompanied by an enrichment of the residual soil N, making it appear that any increase in N uptake by the trees from the fertilization was less than actually occurred (see also Foster *et al.*, 1985).

The utility of further studies of this type for understanding N availability is probably limited. They have helped ascertain the cost effectiveness of fertilizer forms in terms of uptake by trees (Knowles, 1975). However, the problem of isotopic exchange with microbial pools, which is much greater for NH_4^+ than for NO_3^-, adds uncertainty even to these conclusions. The large amount of N added in these studies (typically increasing ambient soil inorganic N pool sizes by several fold) influences N transformations (Johnson *et al.*, 1980), preventing interpretations of native-N cycling within the soil–plant system. An extreme

Table 4. Short-term fate of tracer levels of ^{15}N applied in situ to some forest soils

			Percentage of applied recovered in					
Site/Treatment	^{15}N Applied to	Duration (months)	Plants	Forest floor	Microbial biomass	Solution	SOM	Total
Clearcut[1]	Surface	5.0						
chopped			11	31	12	–	40	94
SPD			13	5	15	–	40	73
Mixed conifer[2] 60-yr-old	A horizon	1.3	–	1	17	1	15	34
Mixed conifer[3] >100-yr-old	A horizon	4.0	–	3	4	1	44	52
		12.0	–	4	5	2	42	53

[1]Data from Vitousek and Matson (1985). The two different treatments refer to contrasting methods of site preparation prior to replanting. Chopping leaves most of the structural organic residue on-site, whereas shearing, piling, and disking removes most of this material as well as most of the forest floor. The forest floor fraction in this table includes large (> 1 cm diameter) organic residue (slash). Soil solution recoveries of the label were not reported.
[2]Data from Schimel (1987). Labeled-N was added to the surface mineral soil by injecting through the forest floor. The soil organic matter (SOM) fraction in this table includes dead roots and detritus. Plant uptake was prevented.
[3]Unpublished data from S. Hart and M. Firestone. Experimental design was similar to Schimel (1987) with ^{15}N injected through the forest floor into the underlying mineral soil. Plant uptake was prevented.

example of the fertilization effect on N transformations comes from Foster *et al.* (1985). They showed that fertilization with 200 kg/ha of urea-N raised the pH of forest floor materials sufficiently to allow chemical incorporation of NH_3 into organic matter in sterilized soils. In unsterilized soils, the pH effect of fertilization allowed chemical immobilization to exceed microbial immobilization.

b. Tracer Applications

Tracing the movement of native-N through the soil–plant system can be achieved using low-level or "tracer-level" additions of ^{15}N (Van Cleve and White, 1980). In general, additions of tracer ^{15}N are kept to less than 50% of the ambient inorganic N pool. However, we know of no study with forest soils that directly examined the potential "fertilizer" effects of these low amounts. This constraint requires the use of highly enriched ^{15}N compounds (usually 70 to 99% atom enrichment).

The experimental approach to tracer ^{15}N techniques is similar to fertilizer-level applications; the amount of the label recovered in various ecosystem compartments is followed over a period of time (Table 4). In the tracer-level case, however, it is assumed that ambient rates of N transformations have been unaltered. Vitousek and Matson (1985) used this approach to show that the forest floor and logging debris left behind following timber harvest were both important for conserving the N capital of the site during reforestation (Table 4). Microbial immobilization associated with these high C/N materials was the most important process in preventing N loss. J.P. Schimel (1987) used low-level additions of ^{15}N

injected in solution through the forest floor into the surface mineral soil to trace N movement in a mature mixed conifer forest in California (Table 4). He found that dead roots and detritus were important sinks for inorganic N in the mineral soil, as was microbial biomass associated with the more decomposed SOM. Little N moved upward into the overlying forest floor over a 1.3 month period. In a similar study, Hart and Firestone (unpublished data) found there was little net ^{15}N movement among soil pools between 4 and 12 months following label addition (Table 4). Apparently, the fate of N recently added to the inorganic N pool is determined fairly rapidly in these forest soils. Weber and Van Cleve (1984) applied tracer amounts of $^{15}NH_4Cl$ and $K^{15}NO_3$ to the forest floors of two black spruce (*Picea mariana* (Mill. B.S.P.) ecosystems in interior Alaska. Using repeated sampling of various ecosystem components over a 2+-year period, they demonstrated that the forest floor acted as a significant nutrient sink, reducing N loss from the ecosystem. Further, they found that both NH_4^+ and NO_3^- were rapidly converted from available to unavailable forms within the forest floor.

Even with several sampling periods, the utility of these short-term "fate" studies is limited because the dynamic nature of N movement in the soil cannot be traced directly. The usefulness of such studies is greatly enhanced by attempting to estimate gross rates of N processes by using the techniques described next. Quantification of the gross rates of N transformations in addition to the longer-term net flows allows the factors that control N availability in forest soils to be elucidated.

3. Estimating Gross N Transformation Rates

The addition of low levels of ^{15}N into a pool allows the simultaneous transfers into and out of the labeled pool to be estimated. This method, generically called pool dilution, has been used primarily only in agricultural soils until fairly recently (Jansson, 1971). Although the basic procedure of adding small amounts of highly enriched $^{15}NH_4^+$ or $^{15}NO_3^-$ is similar among studies, the analysis of the pool dilution data can take many forms.

In a classic study that was one of the first applications of the ^{15}N pool dilution approach to a forest system, Van Cleve and White (1980) characterized the dynamics of NH_4^+, NO_3^-, and soluble and nonsoluble organic-N pools in a 60-year-old paper birch forest floor in interior Alaska. Estimates of fluxes through and turnover times of these various pools were made assuming steady-state conditions (i.e., the size of each pool was assumed constant) and multi-exponential analysis of ^{15}N isotope dilution curves (Rescigno and Segre, 1966; Shipley and Clark, 1972). Among their more novel findings was that fluxes into and out of the soluble organic-N pool were very large, ranging from 0.4 to 0.7 kg-N/ha daily in this boreal forest.

A variety of mathematical models have been used to characterize the transfer among pools in ^{15}N experiments, starting with Kirkham and Bartholomew (1954, 1955). These authors viewed the soil as a two-compartment system of available and unavailable N atoms. After addition of $^{15}NH_4^+$ to the soil, total NH_4^+ and

$^{15}NH_4^+$ are measured in a KCl extract. Calculation of gross rates into and out of the NH_4^+ pool requires only the knowledge of the initial and final total NH_4^+ and $^{15}NH_4^+$ pool sizes. Estimation of gross nitrification and NO_3^- consumption is determined in a similar manner with $^{15}NO_3^-$ additions. In Kirkham and Bartholomew's analysis, two general cases are considered: (1) input and output rates of the labeled pool are constant over the incubation period, (2) rates of N flux are proportional to the size of the donor pool (so-called donor-controlled fluxes). For short time periods (about 1 day) Case 1 is likely applicable (Kirkham and Bartholomew, 1955). A major shortcoming of this technique is its sensitivity to the estimation of the initial size of the labeled pool. As described earlier, the spatial variability in the sizes of NH_4^+ and NO_3^- pools in forest soils is usually large, severely limiting the precision of this method when applied in the field. Further, a "time zero" extraction of the soil immediately after the label is added to the soil is preferred for estimating the initial ^{15}N atom percentage enrichment of the labeled soil pool rather than simply calculating the value from the measured total N content of the pool and the amount of ^{14}N and ^{15}N added.

Because only one input and output rate is calculated for the labeled pool using Kirkham and Bartholomew's equations, estimation of gross microbial immobilization of NH_4^+ or NO_3^- is difficult using this technique. Other possible fates for NH_4^+ include nitrification, leaching, and chemical fixation. Closed systems of incubation can prevent leaching losses, and nitrification can be inhibited by the addition of C_2H_2 (Nishio *et al.*, 1985). However, chemical fixation of NH_4^+ can be substantial in mineral or organic horizons (Jansson, 1971; Foster *et al.*, 1985). Similar difficulty is found when estimating NO_3^- microbial immobilization because other fates for NO_3^- also may exist (i.e., leaching and denitrification).

Nitrogen immobilization may also be estimated for short-term incubations simply from the amount of ^{15}N excess found in the organic-N pool and dividing this value by the mean fractional enrichment of $^{15}NH_4^+$ or $^{15}NO_3^-$ during the incubation period (Shen *et al.*, 1984). As in the above technique, remineralization of immobilized N is assumed negligible. In this case, the total organic N pool is assessed, so highly enriched ^{15}N salts must be used and substantial immobilization must occur for a detectable increase in the organic-^{15}N fraction. Further, chemical immobilization of ^{15}N by nonmicrobial biomass organic materials also will be included in the estimate.

For longer term ($>$ 24 hours) incubations, compartment, or "box" models are used, where the various compartments (NH_4^+, NO_3^-, microbial biomass, etc.) are connected by flows (mineralization, immobilization, etc.) between them. Using the mass balance conditions for ^{15}N and total N in each compartment, a set of differential or finite-difference equations are produced. These are then solved analytically (Nishio *et al.*, 1985), iteratively (Vitousek and Andariese, 1986) or using statistical analysis (Myrold and Tiedje, 1986). The advantage of this later technique is that it includes no assumption about the kinetic order of the N transformations. Examples of the use of these compartment models for estimating simultaneous gross N transfer rates in soil include: Paul and Juma, 1981; Labrouse *et al.*, 1985; Nishio *et al.*, 1985; Myrold and Tiedje, 1986; Vitousek

Table 5. Gross and net N mineralization and nitrification rates for some forest soils

Site	Soil type	Depth or horizon	Mineralization Gross	Mineralization Net	Nitrification Gross	Nitrification Net
			μg N·g soil^{-1}·d^{-1}			
Pinus taeda[1] 22-yr-old	Ultisol	0–15 cm	1.2	0.15	nd	0.0
Northern[2] hardwoods	Spodosol	surface mineral	1.3	0.7	nd	0.5
Mixed conifer[3] 60-yr-old	Alfisol	02	130	1.6	nd	0.17
		0–5 cm	25	0.30	nd	0.06
Mixed conifer[4] 10-yr-old	Alfisol	0–9 cm	0.67–1.7	−0.02–0.13	0.03–1.1	−0.04–0.14
Mixed conifer[4] >100-yr-old	Alfisol	02	20–41	0.14–1.14	2.5–15	−0.02–1.5
		0–9 cm	2.0–5.7	−0.02–0.10	0.23–0.97	−0.07–0.06

[1]Calculated from 29-day laboratory incubation data of Vitousek and Andariese (1986) using initial inorganic N pool sizes and assuming first-order kinetics. N transfer coefficients calculated by the authors using a linear donor-controlled model.
[2]Data from Myrold and Tiedje (1986). Gross mineralization was calculated using a 21-day laboratory incubation and nonlinear regression techniques. Gross nitrification was assumed to be equal to net nitrification.
[3]Gross mineralization determined for 1-day in-field tracer experiments using a pool dilution model (Schimel, 1987). Net rates also determined in-the-field over a 39-day period.
[4]Gross rates determined for 1-day in-field tracer experiments using a pool dilution model. Net rates also determined in-the-field over a 30-day period. Ranges in rates provided reflect seasonal variation (E. Davidson, S. Hart, J. Stark, and M. Firestone, unpublished data).

and Andariese, 1986; and Barraclough and Smith, 1987. Only two of these studies used forest soils, and all but one of them relied on laboratory (using sieved and moistened soils) rather than in-field incubation experiments. This group of studies use longer incubation periods (several weeks) during which transfer rates cannot be assumed constant and several cycles of mineralization and immobilization may occur. However, the models used to simulate transfer of labeled and unlabeled N in the soil–plant system are quite diverse, and different models may fit experimental data equally well (Myrold and Tiedje, 1986). Because N transfer rates estimated by these various models are dependent on the assumptions made in model development, very different estimates of gross transfer rates may be obtained with two different models using the same tracer data set (see Labrouse *et al.*, 1985; D.S. Schimel and Innis, 1986). With these limitations in mind, studies of gross mineralization rates in forest soils (Table 5) clearly show the importance of immobilization in reducing the amount of mineralization that is apparent as net mineralization. The ratio of gross/net mineralization ranged from about 2 to >100, demonstrating that "net mineralization" measures are tremendously affected by the degree of immobilization. Any treatment of a forest ecosystem may alter the balance between gross and net mineralization, and each method of assessing available N may also alter the balance. Very little information is available on shifts in the balance between gross and net mineralization in forest ecosystems, and this area of research is one of the most important topics of current research on N availability. It is clear from this discussion that great opportunities remain for exploiting the power of ^{15}N tracer studies in forest soils. Future efforts on quantification of field rates of gross transfers need to be made using in-field labeling techniques. Further, studies on the controls on N transformations should be conducted in both the laboratory and field, with and without additions of various C and N substrates and chemicals that block specific transformations.

IV. Conclusions

No method provides a clear, accurate assessment of the nutrient supply rate in a soil, or of the degree to which the supply rate limits tree growth. The selection of the appropriate method for each study needs to be based on the objectives of the study, especially on the components of the N cycle (Fig. 1) that are expected to be of interest. Large regional studies may find that soil total N is an adequate index of N availability (see Post *et al.*, 1985), whereas local studies of the effects of stand disturbance require a method that is sensitive to on-site environmental conditions (Table 6).

The quantity of NH_4^+ and NO_3^- extractable at a single point in time is unlikely to provide clear information on N availability. We recommend that this measure be used only in providing an estimate of the initial quantity of NH_4^+ and NO_3^- present at the start of incubations and that no strong inferences be attempted without further information.

The chemical extraction of labile pools of N may provide a useful index of N availability in some cases. These methods do not allow for microbial immobilization, and they may be most useful for comparing soils with very low net mineralization rates that result from high rates of microbial immobilization.

The quantity of total N present in soils may be useful for broad, regional comparisons where differences in microenvironment and substrate quality are overshadowed by macroclimatic effects. Total N analyses may also be informative supplements to other measures of N availability; the quantity mineralized can be expressed as a proportion of the total N present (rather than a proportion of soil weight) to give an index of substrate quality.

Aerobic and anaerobic incubations in the laboratory provide moderately useful information on N mineralization under standard conditions. Although the standardized conditions are artificial, they provide an opportunity for comparisons across sites and across studies that would not be possible under varying conditions.

Repeated incubations provide an opportunity to examine longer-term mineralization, and the removal of the products of mineralization (by periodic leaching) helps reduce the artificialities associated with the accumulation of NH_4^+ and NO_3^-. These assays are limited by the lack of fresh inputs of detritus; the large

Table 6. Summary of the key features of methods for assessment of nitrogen availability

Method	Temperature	Moisture	Immobilization	Nitrification	Plant uptake	Product accumulation	Common units	Effort	Information
Extractable NH_4^+ + NO_3^-	NA	NA	NA	~Yes	No	No	μg N/g soil	Low	Very low
Chemical extraction of labile pools	NA	NA	NA	No	No	No	μg N/g soil	Low	Low to moderate
Total N	NA	NA	NA	No	No	No	%N, μg N/g soil	Moderate	Low to moderate
Aerobic incubations	Constant	Constant	Important	Yes	No	Yes	μg N/g soil/time	Moderate	Moderate
Anaerobic incubations	Constant	Constant	Important?	No	No	Yes	μg N/g soil/time	Low	Moderate
Repeated incubations	Constant	Constant	Important	Yes	No	No	μg N/g soil/time; N_0	High	Moderate to high
Microbial biomass N	Constant	Constant	Important	No	No	Yes	μg N/g soil	High	Moderate to high
Bioassay	Variable	Variable	Important	~Yes	Yes	No	μg N/g soil; μg N/pot	High	Moderate
Ecosystem N budget	Variable	Variable	Important	~No	Yes	No	g N/m²/yr; kg N/ha/yr	Very high	High
Ion exchange resin bags	Variable	Variable	Important	Yes	Yes	No	μg N/g resin; μg N/bag	Low	Moderate
Buried-bag incubation	Variable	Constant	Important	Yes	No	Yes	μg N/g soil/time	Moderate	Moderate to high
Closed-top core	Variable	~Constant	Important	Yes	No	Yes	μg N/g soil/time	Moderate	Moderate to high
Resin-core incubation	Variable	Variable	Important	Yes	No	No	μg N/g soil/time	High	Moderate to high

[a] The columns for temperature and moisture indicate whether variations in these factors are included in the methods.
[b] The columns for immobilization, nitrification, plant uptake, and product accumulation indicate whether these processes are important during the assays, but they do not reflect the possible role prior to the assay.
[c] The columns for effort and information provide a relative scale for comparing the time and cost of the assays with the probable value of the information obtained.

seasonal differences in N mineralization and N_0 found by this method in forest soils emphasize the importance of these inputs.

Assays of microbial biomass N are poorly developed indexes of available N in forest soils, but further developments promise to provide both useful availability estimates and insight into the processes underlying estimates obtained by other assays. Additional work is especially needed to identify the flux of N through the microbial pool as well as the importances of major taxonomic groups of microbes.

Bioassays may provide useful estimates of N availability, if the plant species used grows rapidly enough and if the importances of soil structure and microenvironment are slight relative to substrate quality. Bioassays may be most useful when combined with other measures of N availability.

The N budget of ecosystems remains the best measure of the net availability of N to trees, but this intensive approach provides little insights into the components of N mineralization. Quantification of the net fluxes of N within an ecosystem provides an excellent framework for the evaluation of process-oriented measures of N availability.

The accumulation of NH_4^+ and NO_3^- in ion exchange resin bags is regulated by the same processes that affect the supply of these ions to root surfaces: mineralization, immobilization, and ion transport. However, the relative importances of these factors may differ between resin bags and root surfaces, and the interpretation of resin bag values can be problematic. Resin bags may be most useful for examining spatial and temporal patterns of N availability owing to ease-of-use and high sensitivity to important factors. Some laboratories have had difficulties analyzing resin extracts; readers interested in this method need to contact others who have used it and to employ quality assurance measures.

Buried-bag incubations have become the standard method for ecosystem ecologists assessing N availability in North American forests. They require more information than laboratory assays, but the increased value of the information obtained generally justifies the work. Given the widespread use of this method, it is especially important that more research focus on the specific components that determine the net mineralization values. Experimental manipulations of temperature, moisture, incubation period, and soil disturbances (in combination with ^{15}N tracer techniques) are needed to evaluate the importance of these factors on the net mineralization estimate.

Closed-top cores offer most of the advantages of buried bags, but with less soil disturbance. The possibility of nitrate leaching from the bottoms of the cores does not appear to have been important in earlier studies, but this should probably be verified for other types of forests in the future. We recommend that studies using buried bags also use closed-top cores; if the closed-top cores compare favorably, they might be adopted for future use.

Resin core incubations offer potential advantages over buried-bag and closed-top core incubations, but they take significantly more effort. The advantage of allowing precipitation inputs to recharge the soil core (and to leach ions from the core) may be unimportant if the timing of core replacement matches precipitation events. The hydraulic discontinuity between the resin bag and the soil may keep the moisture content in the cores marginally above the content of the surrounding soil; closed-top tubes may be more effective in matching ambient soil moisture levels. We currently recommend resin cores for high-precipitation environments or when the length of time between collections exceeds one month.

All the methods described here need more research on the contribution of each component in Figure 1 to the estimate of available N. Coupling net N mineralization with C mineralization during incubation studies provides a first step toward identifying the key components. Nitrogen tracer experiments provide great opportunities for unraveling the components of the N availability estimates; the ease-of-use and low cost of ^{15}N techniques should lead to a great expansion in this field. Despite the decades of research that have been invested in assessing N availability in forests, we feel that considerable work remains in identifying the importance of the components of the N cycle to each method and to identifying the most appropriate method for specific research objectives.

Acknowledgments

This review and synthesis developed from discussions with John Pastor; we thank him for his ideas and for helpful comments on the manuscript. We thank David Valentine, Knute Nadelhoffer, Bob Powers, and John Aber for their help in improving the manuscript. Funding for this paper was provided by NSF grants BSR-8416768 and BSR-8306181, and BSR-8807771, and McIntire-Stennis Projects 0133245 (CSU) and 4262 MS (Univ. of California).

References

Aber, J., J.M. Melillo, K. Nadelhoffer, C. McClaugherty, and J. Pastor. 1985. Fine root turnover in forest ecosystems in relation to quantity and form of nitrogen availability: A comparison of two methods. *Oecologia* 66:317–321.

Adams, M., and P. Attiwill. 1986a. Nutrient cycling and nitrogen mineralization in eucalypt forests of south-eastern Australia. I. Nutrient cycling and nitrogen turnover. *Plant and Soil* 92:319–339.

Adams, M., and P. Attiwill. 1986b. Nutrient cycling and nitrogen mineralization in eucalypt forests of south-eastern Australia. II. Indices of nitrogen mineralization. *Plant and Soil* 92:341–362.

Amato, M., and J.N. Ladd. 1980. Studies of nitrogen immobilization and mineralization in calcareous soils—V. Formation and distribution of isotope-labelled biomass during decomposition of ^{14}C- and ^{15}N-labelled plant material. *Soil Biol. Biochem.* 12:405–411.

Amer, F., D. Bouldin, C. Black, and F. Duke. 1955. Characterization of soil phosphorus by anion exchange resin adsorption and P-32 equilibration. *Plant and Soil* 6:391–408.

Antonietti, A. 1968. Le associazioni forestali dell'orizzonte submontano del Cantone Ticino su substrati pedogenticic ricchi di carbonati. *Mitt. Schweiz. Anst. Forstl. Versuchswes* 44:85–226.

Azam F., K.A. Malik, and F. Hussain. 1986. Microbial biomass and mineralization-immobilization of nitrogen in some agricultural soils. *Biol. Fert. Soil* 2:157–163.

Barraclough, D., and M.J. Smith. 1987. The estimation of mineralization, immobilization and nitrification in nitrogen-15 field experiments using computer simulation. *J. Soil Sci.* 38:519–530.

Barber, S.A. 1974. Influence of the plant root on ion movement in soil. In: E.W. Carson (ed.), *The plant root and its environment.* Univ. Press of Virginia, Charlotte, pp. 525–564.

Binkley, D. 1982. Nitrogen fixation and net primary production in a young Sitka alder stand. *Can. J. Bot.* 60:281–284.

Binkley, D. 1983. Ecosystem production in Douglas-fir plantation: Interaction of red alder and site fertility. *For. Ecol. Manage.* 5:215–227.

Binkley, D. 1984a. Does forest removal increase rates of decomposition and nitrogen availability? *For. Ecol. Manage.* 8:229–233.

Binkley, D. 1984b. Ion exchange resin bags: Factors affecting estimates of nitrogen availability. *Soil Sci. Soc. Am. J.* 48:1181–1184.

Binkley, D. 1986. *Forest nutrition management.* Wiley, NY. 290 pp.

Binkley, D., and P. Matson. 1983. Ion exchange resin bag method for assessing forest soil nitrogen availability. *Soil Sci. Soc. Am. J.* 47:1050–1052.

Binkley, D., and P. Vitousek. 1989. Soil nutrient availability. In: R. Pearcy, H. Mooney, J. Ehleringer, and P. Rundel (eds.), *Physiological plant ecology: Field methods and instrumentation.* Chapman & Hall, London, pp. 75–96.

Binkley, D., K. Cromack, Jr., and R.L. Fredriksen. 1982. Nitrogen accretion and availability in some snowbrush ecosystems. *For. Sci.* 28:720–724.

Binkley, D., P. Sollins, and W. McGill. 1985. Natural abundance of nitrogen-15 as a tool for tracing alder-fixed nitrogen. *Soil Sci. Soc. Am. J.* 49:444–447.

Binkley, D., J. Aber, J. Pastor, and K. Nadelhoffer. 1986. Nitrogen availability in some Wisconsin forests: Comparisons of resin bags and on-site incubations. *Biol. and Fert. of Soils* 2:77–82.

Binkley, D., P. Sollins, R. Bell, D. Sachs, and C. Glassman. *In review.* Biogeochemistry of adjacent conifer and alder/conifer ecosystems. Submitted to *Ecology.*

Birch, H.F. 1958. The effect of soil drying on humus decomposition and nitrogen availability. *Plant and Soil* 10:9–31.

Birch, H.F. 1960. Nitrification in soils after different periods of dryness. *Plant and Soil* 12:81–96.

Bloss, S., and D. Binkley. *In review.* Effect of rooting by wild boars on nitrogen mineralization in high elevation beech forests of the southern Appalachians. Submitted to *Can. J. For. Res.*

Bonde, T.A., and T. Rosswall. 1987. Seasonal variation of potentially mineralizable nitrogen in four cropping systems. *Soil Sci. Soc. Am. J.* 51:1508–1514.

Bremner, J. 1965. Nitrogen availability index. In. C. Black (ed.), *Methods of soil analysis, part 2.* Am. Soc. Agron., Madison, WI, pp. 1324–1345.

Bremner, J., and C. Mulvaney. 1982. Nitrogen—total. In: A. Page, R. Miller, and D. Keeney (eds.), *Methods of soil analysis, part 2, chemical and microbiological properties.* Am Soc. Agron., Madison, WI, pp. 595–624.

Brookes, P.C., A. Landman, G. Pruden, and D.S. Jenkinson. 1985. Chloroform fumigation and the release of soil nitrogen: A rapid direct extraction method to measure microbial biomass nitrogen in soil. *Soil Biol. Biochem.* 17:837–842.

Burger, J., and W. Pritchett. 1984. Effects of clearfelling and site preparation on nitrogen mineralization in a Southern pine stand. *Soil Sci. Soc. Am. J.* 48:1432–1437.

Carter, M.R. 1986. Microbial biomass and mineralizable nitrogen in solonetzic soils: Influence of gypsum and lime amendments. *Soil Biol. Biochem.* 18:531–537.
Carter, M.R., and J.A. Macleod. 1987. Biological properties of some Prince Edward Island soils: Relationship between microbial biomass nitrogen and mineralizable nitrogen. *Can. J. Soil Sci.* 67:333–340.
Carter, M.R., and D.A. Rennie. 1982. Changes in soil quality under zero tillage farming systems: distribution of microbial biomass and mineralizable C and N potentials. *Can. J. Soil Sci.* 62:587–597.
Chapin, F.S.H. III, K. Van Cleve, and P. Vitousek. 1986. The nature of nutrient limitation in plant communities. *Am. Naturalist* 127:148–158.
Christensen, N., and T. MacAller. 1985. Soil mineral nitrogen transformations during succession in the Piedmont of North Carolina. *Soil Biol. Biochem.* 17:675–681.
Covington, W., and S. Sackett. 1986. Effect of periodic burning on soil nitrogen concentrations in ponderosa pine. *Soil Sci. Soc. Am. J.* 50:452–457.
Deans, J.R., J. Molina, and C. Clapp. 1986. Models for predicting potentially mineralizable nitrogen and decomposition rate constants. *Soil Sci. Soc. Am. J.* 50:323–326.
Di Stefano, J. 1984. Nitrogen mineralization and non-symbiotic nitrogen fixation in an age sequence of slash pine plantations in north Florida. Ph.D. diss., Univ. of Florida, Gainesville, pp. 219.
Di Stefano, J., and H. Gholz. 1986. A proposed use of ion exchange resin to measure nitrogen mineralization and nitrification in intact soil cores. *Comm. Soil Sci. Plant Anal.* 17:989–998.
Ellenberg, H. 1977. Stickstoff als Standortsfaktor, insbesondere für mitteleuropäische Pflanzengesellschaften. *Oecologia Plantarum* 12:1–22.
Eno, C. 1960. Nitrate production in the field by incubating the soil in polyethylene bags. *Soil Sci. Soc. Am. Proc.* 24:277–279.
Fahey, T.J., J. Yavitt, J. Pearson, and D. Knight. 1985. The nitrogen cycle in lodgepole pine forests, southeastern Wyoming. *Biogeochem.* 1:257–275.
Federer, C.A. 1983. Nitrogen mineralization and nitrification: Depth variation in four New England forest soils. *Soil Sci. Soc. Am. J.* 47:1008–1014.
Feigin, A., D. Kohl, G. Shearer, and B. Commoner. 1974. Variation in the natural nitrogen-15 abundance in nitrate mineralized during incubation of several Illinois soils. *Soil Sci. Soc. Am. Proc.* 38:90–95.
Fisher, R., and E. Stone. 1969. Increased availability of nitrogen and phosphorus in the root zone of conifers. *Soil Sci. Soc. Am. Proc.* 33:955–961.
Flanagan, P., and K. Van Cleve. 1983. Nutrient cycling in relation to decomposition and organic matter quality in taiga ecosystems. *Can. J. For. Res.* 13:795–817.
Foster, N., E. Beauchamp, and C. Corke. 1985. Immobilization of nitrogen-15 labelled urea in a jack pine forest floor. *Soil Sci. Soc. Am. J.* 49:448–452.
Fox, T., J. Burger, and R. Kreh. 1986. Effects of site preparation on nitrogen dynamics in the southern Piedmont. *For. Ecol. Manage.* 15:241–256.
Frazer, D., J. McColl, and R. Powers. *In review.* Mineralization of soil nitrogen in a managed, mixed-conifer forest in northern California. Submitted to *Soil Sci. Soc. Am. J.*
Fyles, J.W., and W.B. McGill. 1987. Nitrogen mineralization in forest profiles from central Alberta. *Can. J. For. Res.* 17:242–249.
Geist, J.M. 1977. Nitrogen response relationship of some volcanic ash soils. *Soil Sci. Soc. Am. J.* 41:996–1000.

Gianello, C., and J. Bremner. 1986. Comparison of chemical methods of assessing potentially available organic nitrogen in soil. *Comm. Soil Sci. Plant Anal.* 17:215–236.

Gibson, D., I. Colquhoun, and P. Greig-Smith. 1985. A new method for measuring nutrient supply rates in soils using ion-exchange resins. In: A.H. Fitter (ed.), *Ecological interactions in soils: Plants, microbes, and animals.* Blackwell, Oxford, pp. 73–79.

Gilliam, F., and D. Richter. 1985. Increases in extractable ions in infertile aquults caused by sample preparation. *Soil Sci. Soc. Am. J.* 49:1576–1578.

Glavac, V., and H. Koenies. 1978a. Mineralstickstoff-Gehalte und N-Nettomineralisation im Boden eines Fictenforstes und seines Kahlschlages während der Vegetationsperiode 1977. *Oecologia Plantarum* 13:207–218.

Glavac, V., and H. Koenies. 1978b. Vergleiche der N-Nettomineralisation in einem Sauerhumus-Buchenwald (Luzulo-Fagetum) und einem benachbarten Fictenforst am gleichen Standort vor und nach dem Kahlschlag. *Oecologia Plantarum* 13:219–226.

Gordon, A., and K. Van Cleve. 1983. Seasonal patterns of nitrogen mineralization following harvesting in the white spruce forests of interior Alaska. In: R.W. Wein, R.R. Riewe, and I.R. Methven (eds.), *Resources and dynamics of the boreal zone.* Assoc. Can. Univ. North. Stud., Ottawa, pp. 119–130.

Gordon, A., M. Tallas, and K. Van Cleve. 1987. Soil incubations in polyethylene bags: Effect of bag thickness and temperature on nitrogen transformations and CO_2 permeability. *Can. J. Soil Sci.* 67:65–75.

Gosz, J.R., and C.S. White. 1986. Seasonal and annual variation in nitrogen mineralization and nitrification along an elevational gradient in New Mexico. *Biogeochem.* 2:281–297.

Harmer, R., and I. Alexander. 1985. Effects of root exclusion on nitrogen transformations and decomposition processes in spruce humus. In: A. Fitter (ed.), *Ecological interactions in soil: Plants, microbes and animals.* Blackwell, Oxford, pp. 267–277.

Harmer, R., and I. Alexander. 1986. The effect of starch amendment on nitrogen mineralisation from the forest floor beneath a range of conifers. *Forestry* 59:39–46.

Hart, S.C. 1988. Carbon and nitrogen accretion and dynamics in volcanic ash deposits from different subarctic habitats. *Biol. Fert. Soils. In press.*

Hart, S.C., and D. Binkley. 1985. Correlations among indices of forest soil nutrient availability in fertilized and unfertilized loblolly pine plantations. *Plant and Soil* 85:11–21.

Hart, S.C., and M.K. Firestone. 1989. Evaluation of three *in situ* nitrogen availability assays. In press. *Can. J. For. Res.*

Hart, S.C., and A.J. Gunther. *In-situ* estimate of annual net nitrogen mineralization and nitrification in a subarctic watershed. *Oecologia. In review.*

Hart, S.C., D. Binkley, and R.G. Campbell. 1986. Predicting loblolly pine growth and growth response to fertilization. *Soil Sci. Soc. Am. J.* 50:230–233.

Hauck, R., and J. Bremner. 1976. Use of tracers for soil and fertilizer nitrogen research. *Adv. Agron.* 28:219–266.

Hedman, C., and D. Binkley. 1988. Canopy profiles of some Piedmont hardwood forests. *Can. J. For. Res.* 18:1090–1093.

Heilman, P.E., T.H. Dao, H.H. Cheng, S.R. Webster, and L. Christensen. 1982. Comparison of fall and spring applications of 15N-labeled urea to Douglas-fir: II. Fertilizer nitrogen recovery in trees and soil after 2 years. *Soil Sci. Soc. Am. J.* 46:1300–1304.

Jansson, S.L. 1958. Tracer studies on nitrogen transformations in soil with special attention to mineralization-immobilization relationships. *Ann. Roy. Agr. Coll. Sweden* 24:101–361.

Jansson, S.L. 1971. Use of ^{15}N in studies of soil nitrogen. In: A.D. McLaren and J. Skujins (eds.), *Soil Biochemistry*, vol. 2. Marcel Dekker, NY, pp. 129–166.

Jansson, S.L., and J. Persson. 1982. Mineralization and immobilization of soil nitrogen. In: F.J. Stevenson (ed.), *Nitrogen in Agricultural Soils.* Am. Soc. Agron., Madison, WI, pp. 229–252.

Jenkinson, D.S., and J.N. Ladd. 1981. Microbial biomass in soil: Measurement and turnover. In: E.A. Paul and J.N. Ladd (eds.), *Soil Biochemistry,* vol. 5. Marcel Dekker, NY, pp. 415–471.

Jenkinson, D.S., and D.S. Powlson. 1976. The effects of biocidal treatments on metabolism in soil. V: A method for measuring soil biomass. *Soil Biol. Biochem.* 8:209–213.

Johnson, D.W., N.T. Edwards, and D.E. Todd. 1980. Nitrogen mineralization, immobilization, and nitrification following urea fertilization of a forest soil under field and laboratory conditions. *Soil Sci. Soc. Am. J.* 44:610–616.

Kadeba, O., and J. Boyle. 1978. Evaluation of phosphorus in forest soils: Comparison of phosphorus uptake, extraction method and soil properties. *Plant and Soil* 49:285–297.

Keeney, D. 1980. Prediction of soil nitrogen availability in forest ecosystems: A literature review. *For. Sci.* 26:159–171.

Keeney, D. 1982. Nitrogen-availability indices. In: A. Page, R. Miller, and D. Keeney (eds.), *Methods of soil analysis, part 2, chemical and microbiological properties.* Am. Soc. Agron., Madison, WI, pp. 711–734.

Keeney, D., and J. Bremner. 1966. A chemical index of soil nitrogen availability. *Nature* 211:892–893.

Keeney, D., and D. Nelson. 1982. Nitrogen-inorganic forms. In: A. Page, R. Miller, and D. Keeney (eds.), *Methods of soil analysis, part 2, chemical and microbiological properties.* Am. Soc. Agron., Madison, WI, pp. 643–698.

Kirkham, D., and W.V. Bartholomew. 1954. Equations for following nutrient transformations in soil, utilizing tracer data. *Soil Sci. Soc. Am. Proc.* 18:33–34.

Kirkham, D., and W.V. Bartholomew. 1955. Equations for following nutrient transformations in soil, utilizing tracer data: II. *Soil Sci. Soc. Am. Proc.* 19:189–192.

Knowles, R. 1975. Interpretation of recent ^{15}N studies of nitrogen in forest systems. In: B. Bernier and C.H. Winget (eds.), *Forest soils and forest land management.* Proc. of the 4th N. Am. For. Soil Conf., Laval Univ., Quebec. August 1973. Laval Univ. Press, pp. 53–65.

Kohl, D., G. Shearer, and J. Harper. 1980. Estimates of N_2 fixation based on differences in the natural abundance of ^{15}N in nodulating and nonnodulating isolines of soybeans. *Plant Physiol.* 66:61–65.

Kovacs, M. 1978. Stickstoffverhältnisse im Boden des Eichen-Zerreichen-Walkökosystems. *Oecologia Plantarum* 13:75–82.

Krause, H., and D. Ramlal. 1987. In situ nutrient extraction by resin from forested, clearcut and site-prepared soil. *Can. J. Soil Sci.* 67:943–952.

Kutiel, P., and Z. Naveh. 1987. The effect of fire on nutrients in a pine forest soil. *Plant and Soil* 104:269–274.

Labrouse, L., W. Vidal, J.C. Tosca, and P.H. Berge. 1985. Transferts d'azote mineral dans les sols froids: Essai de tracage isotopique *in situ. Soil Biol. Biochem.* 17:683–689.

Lajtha, K. 1988. The use of ion-exchange resin bags for measuring nutrient availability in an arid ecosystem. *Plant and Soil* 105:105–111.

Lamb, D. 1975. Patterns of nitrogen mineralization in the forest floor of stands of *Pinus radiata* on different soils. *J. Ecol.* 63:615–625.

Lamb, D. 1980. Soil nitrogen mineralisation in a secondary rainforest succession. *Oecologia* (Berl.) 47:257–263.

Lea, R., and R. Ballard. 1982. Predicting loblolly pine growth response from N fertilizer, using soil-N availability indices. *Soil Sci. Soc. Am. J.* 46:1096–1099.

Lemée, G. 1967. Investigations sur la minéralisation de l'azote et son évolution annuelle dans des humus forestiers *in situ*. *Oecologia Plantarum* 2:285–324.

Mahendrappa, M. 1980. Relationships between different estimates of mineralizable N in the organic materials under black spruce stands. *Can. J. For. Res.* 10:517–522.

Mahendrappa, J., N. Foster, G. Weetman, and H. Krause. 1986. Nutrient cycling and availability in forest soils. *Can. J. Soil Sci.* 66:547–572.

Maimone, R., L. Morris, and T. Fox. 1987. Nitrogen mineralization potential in a Lower Coastal Plain forest soil. *Agron. Abstr.* 1987:261.

Marion, G., and C. Black. 1987. The effect of time and temperature on nitrogen mineralization in arctic tundra soils. *Soil Sci. Soc. Am. J.* 51:1501–1508.

Marumoto, T., H. Kai, T. Yoshida, and T. Harada. 1977a. Drying effect on mineralization of microbial cells and their cell walls in soil and contribution of microbial cell walls as a source of decomposable soil organic matter due to drying. *Soil Sci. Plant Nutr.* 23:9–19.

Marumoto, T., H. Kai, T. Yoshida, and T. Harada. 1977b. Relationship between an accumulation of soil organic matter becoming decomposable due to drying of soil and microbial cells. *Soil Sci. Plant Nutr.* 23:1–8.

Marumoto, T., J.P.E. Anderson, and K.H. Domsch. 1982a. Decomposition of ^{14}C- and ^{15}N-labelled microbial cells in soil. *Soil Biol. Biochem.* 14:461–467.

Marumoto, T., J.P.E. Anderson, and K.H. Domsch. 1982b. Mineralization of nutrients from soil microbial biomass. *Soil Biol. Biochem.* 14:469–475.

Matson, P., and R. Boone. 1984. Natural disturbance and nitrogen mineralization: Wave-form dieback of mountain hemlock in the Oregon Cascades. *Ecology* 65:1511–1516.

Matson, P., and P. Vitousek. 1981. Nitrogen mineralization and nitrification potentials following clearcutting in the Hoosier National Forest, Indiana. *For. Sci.* 27:781–791.

Matson, P., P. Vitousek, J. Ewel, M. Mazzarino, and P. Robertson. 1987. Nitrogen transformations following tropical forest felling and burning on a volcanic soil. *Ecology* 68:491–502.

McNabb, D.H. 1984. Handling and storage of soil samples. In: *Nitrogen Assessment Workshop*, Regional Forest Nutrition Research Project Report No. 2, College of Forest Resources, Univ. of Washington, Seattle, pp. 32–34.

Mead, D., and W. Pritchett. 1971. A comparison of tree response to fertilizers in field and pot experiments. *Soil Sci. Soc. Am. Proc.* 35:246–349.

Mead, D.J., and W.L. Pritchett. 1975. Fertilizer movement in a slash pine ecosystem. II. N distribution after two growing seasons. *Plant and Soil* 43:467–478.

Melillo, J. 1977. Mineralization of nitrogen in northern forest ecosystems. Ph.D. diss., Yale Univ. *Diss. Abstr.* 38:3026–B.

Melillo, J. 1981. Nitrogen cycling in deciduous forests. In: F. Clark, and T. Rosswall (eds.), *Terrestrial nitrogen cycles. Ecol. Bull.* (Stockholm) 33:427–442.

Melin, J., H. Nômmik, U. Lohm, and J. Flower-Ellis. 1983. Fertilizer nitrogen budget in a Scots pine ecosystem attained by using root-isolated plots and ^{15}N tracer technique. *Plant and Soil* 74:249–263.

Miller, H., J. Cooper, and J. Miller. 1976. Effect of nitrogen supply on nutrient uptake in a stand of Corsican pine. *J. Appl. Ecol.* 13:955–966.

Mladenoff, D. 1987. Dynamics of nitrogen mineralization and nitrification in hemlock and hardwood treefall gaps. *Ecology* 68:1171–1180.

Montagnini, F., B. Haines, L. Boring, and W. Swank. 1986. Nitrification potentials in early successional black locusts and in mixed hardwood forest stands in the southern Appalachians, USA. *Biogeochem.* 2:197–210.

Myrold, D.D. 1987. Relationship between microbial biomass nitrogen and a nitrogen availability index. *Soil Sci. Soc. Am. J.* 51:1047–1049.

Myrold, D.D., J.M. Tiedje. 1986. Simultaneous estimation of several nitrogen cycle rates using ^{15}N: Theory and application. *Soil Biol. Biochem.* 18:559–568.

Nadelhoffer, K., J. Aber, and J. Melillo. 1983. Leaf-litter production and soil organic matter dynamics along a nitrogen-availability gradient in Southern Wisconsin (U.S.A.). *Can. J. For. Res.* 13:12–21.

Németh, K., I. Makhdum, K. Koch, and H. Beringer. 1979. Determination of categories of soil nitrogen by electro-ultrafiltration (EUF). *Plant and Soil* 53:445–453.

Nishio, T., T. Kanamori, and T. Fujimoto. 1985. Nitrogen transformation in aerobic soil as determined by a $^{15}NH_4^+$ dilution technique. *Soil Biol. Biochem.* 17:149–154.

Nordmeyer, H., and J. Richter. 1985. Incubation experiments on nitrogen mineralization in loess and sandy soils. *Plant and Soil* 83:443–445.

Olson, R., and W. Reiners. 1983. Nitrification in subalpine balsam fir soils: Tests for inhibitory factors. *Soil Biol. Biochem.* 15:413–418.

Pastor, J., and W.M. Post. 1986. Influence of climate, soil moisture, and succession on forest carbon and nitrogen cycles. Biogeochem 2:3–27.

Pastor, J., and W.M. Post. 1988. Response of northern forests to CO_2-induced climate change. *Nature. In press.*

Pastor, J., M.A. Stillwell, and D. Tilman. 1987a. Nitrogen mineralization and nitrification in four Minnesota old fields. *Oecologia* 71:481–485.

Pastor, J., J. Aber, C. McClaugherty, and J. Melillo. 1984. Aboveground production and N and P cycling along a nitrogen mineralization gradient on Blackhawk Island, Wisconsin. *Ecology* 65:256–268.

Pastor, J., R. Garner, V. Dale, and W. Post. 1987b. Successional changes in nitrogen availability as a potential factor contributing to spruce declines in boreal North America. *Can. J. For. Res.* 17:1394–1400.

Paul, E.A., and N.G. Juma. 1981. Mineralization and immobilization of soil nitrogen by microorganisms. In: F.E. Clark and T. Rosswall (eds.), *Terrestrial nitrogen cycles. Ecol. Bull.* (Stockholm), 33:179–195.

Paul, E.A., J.L. Smith, and J.M. Norton. 1986. The decomposition of ^{14}C- and ^{15}N-labeled cells in soil under anaerobic conditions. *Agron. Abst.* 1986:186.

Peterson, B., and B. Fry. 1987. Stable isotopes in ecosystem studies. *Ann. Rev. Ecol. Syst.* 18:293–320.

Peterson, C., P. Ryan, and S. Gessel. 1984. Response of Northwest Douglas-fir stands to urea: Correlations with forest soil properties. *Soil Sci. Soc. Am. J.* 48:162–169.

Pfadenhauer, J. 1979. Die Stickstoffmineralisation in Böden subtropischer Regenwälder in Südbrasilien. *Oecologia Plantarum* 14:27–40.

Poovarodom, S., R. Tate, and R. Bloom. 1988. Nitrogen mineralization rates of the acidic, xeric soils of the New Jersey pinelands: Field rates. *Soil Science* 145:257–263.

Post, W.M., J. Pastor, P. Zinke, and A. Stangenberger. 1985. Global patterns of soil nitrogen storage. *Nature* 317:613–616.

Powers, R. 1980. Mineralizable soil nitrogen as an index of nitrogen availability to forest trees. *Soil Sci. Soc. Am. J.* 44:1314–1320.

Powers, R. 1984a. Estimating soil nitrogen availability through soil and foliar analysis. In: E. Stone (ed.), *Forest soils and treatment impacts.* Proc. of the 6th N. Am. For. Soils Conf. Univ. of Tennessee, Knoxville, August 1983, pp 353–379.

Powers, R.F. 1984b. Site productivity and soil nitrogen status. In: *Nitrogen assessment workshop*, Regional Forest Nutrition Research Project Report No. 2, College of Forest Resources, Univ. of Washington, Seattle, pp. 60–68.

Powers, R.F. 1984c. Sources of variation in mineralizable soil nitrogen. In: *Nitrogen assessment workshop*, Regional Forest Nutrition Research Project Report No. 2, College of Forest Resources, Univ. of Washington, Seattle, pp. 25–31.

Powers, R. 1987. Soil nitrogen mineralization along an altitudinal gradient: Influence of soil temperature and moisture. Presented in IUFRO Conf. on Management of Water and Nutrient Relations to Increase Forest Growth, Canberra, ACT, October 19–22, 1987.

Radwan, M., and J. Shumway. 1983. Soil nitrogen, sulfur, and phosphorus in relation to growth response of western hemlock to nitrogen fertilization. *For. Sci.* 29:469–477.

Raison, R., M. Connell, and P. Khanna. 1987a. Methodology for studying fluxes of soil mineral-N *in situ*. *Soil Biol. Biochem.* 19:521–530.

Raison, R., P. Khanna, M. Connell, and R. Falkiner. 1987b. Effects of water supply and fertilization on N cycling in a stand of *Pinus radiata*. Presented at IUFRO Conf. on Management of Water and Nutrient Relations to Increase Forest Growth, Canberra, ACT, October 19–22, 1987.

Rapp, M., M. Leclerc, and P. Lossaint. 1979. The nitrogen economy in a *Pinus pinea* L. stand. *For. Ecol. Manage.* 2:221–231.

Rashid, G. 1987. Effects of fire on soil carbon and nitrogen in a Mediterranean oak forest of Algeria. *Plant and Soil* 103:89–93.

Rescigno, A., and G. Segre. 1966. *Drug and Tracer Kinetics*. Blaisdell, Waltham, MS, 209 pp.

Rice, E., and S. Pancholy. 1972. Inhibition of nitrification by climax ecosystems. *Am. J. Bot.* 59:1033–1040.

Robertson, G.P. 1987. Geostatistics in ecology: Interpolating with known variance. *Ecology* 68:744–748.

Robertson, G.P., and P. Vitousek. 1981. Nitrification potentials in primary and secondary succession. *Ecology* 62:376–386.

Robertson, G.P., M.A. Houston, F.C. Evans, and J.M. Tiedje. 1988. Spatial variability in a successional plant community: Patterns of nitrogen availability. *Ecology* 69:1517–1524.

Runge, M. 1971. Investigations of the content and the production of mineral nitrogen in soils. In: H. Ellenberg (ed.), *Integrated experimental ecology: Methods and results of ecosystem research in the German Solling Project.* Springer-Verlag, NY, pp. 191–202.

Runge, M. 1974. Die Stickstoff-Mineralisation im Boden eines Sauerhumus-Buchenwaldes. I. Mineralstickstoff-Gehalt und Netto-Mineralisation. *Oceologia Plantarum* 9:201–218.

Ryan, M., and W. Covington. 1986. Effect of a prescribed burn in ponderosa pine on inorganic nitrogen concentrations of mineral soil. USDA For. Ser. Res. Note RM-464, Ft. Collins, CO

Safford, L. 1982. Correlation of greenhouse bioassay with field response to fertilizer by paper birch. *Plant and Soil* 64:167–176.

Sahrawat, K.L. 1980. Soil and fertilizer nitrogen transformations under alternate flooding and drying moisture regimes. *Plant and Soil* 55:225–233.

Sasser, C., and D. Binkley. 1988. Nitrogen mineralization in high elevation forests of the Appalachians. II. Patterns with stand development in fir waves. *Biogeochem.* in press.

Satterson, K. 1985. Nitrogen availability, primary production, and nutrient cycling during secondary succession in North Carolina Piedmont forests. Ph.D. diss., Univ. of North Carolina, Chapel Hill, pp. 57–59.

Scarsbrook, C.E. 1965. Nitrogen availability. In: W.V. Bartholomew and F.E. Clark (eds.), *Soil nitrogen agronomy*, Am. Soc. Agron., Madison, WI, 10:481–502.
Schimel, D.S., and G. Innis. 1986. Quantification of nitrogen turnover: Models of isotope dilution in soils. *Ecol. Soc. Am. Abst.* 1986:300.
Schimel, J.P. 1987. Plant/microbial competition for nitrogen in California forest and grassland. Ph.D. diss., Univ. of California, Berkeley, 156 pp.
Seneviratne, R., and A. Wild. 1985. Effect of mild drying on the mineralization of soil nitrogen. *Plant and Soil* 84:175–179.
Shen, S.M., G. Pruden, and D.S. Jenkinson. 1984. Mineralization and immobilization of nitrogen infumigated soil and the measurement of microbial biomass nitrogen. *Soil Biol. Biochem.* 16:437–444.
Shipley, R.A., and R.E. Clark. 1972. *Tracer methods for in vivo kinetics.* Academic Press, NY, 239 pp.
Shumway, J. 1984. Total nitrogen, mineralizable nitrogen and site index as guides to fertilization of Douglas-fir. In: *Nitrogen assessment workshop*, May 19–20, 1982, RFNRP Report No. 2, College of Forest Resources, Univ. of Washington, Seattle, p. 57–59.
Shumway, J., and W.A. Atkinson. 1978. Predicting nitrogen fertilizer response in unthinned stands of Douglas-fir. *Comm. Soil Sci. Plant Anal.* 9:529–539.
Sibbeson, E. 1977. A simple ion-exchange resin procedure for extracting plant-available elements from soil. *Plant and Soil* 46:665–669.
Smith, J.L., J. Norton, E.A. Paul. 1985. Microbial biomass estimations. In: J.L. Smith and E.A. Paul (eds.), *Process controls and nitrogen transformations in terrestrial ecosystems*, Dept. of Plant and Soil Biology, Univ. of California, Berkeley, pp. 94–108.
Smith, J.L., B.L. McNeal, E.J. Owens, and G.O. Klock. 1981. Comparison of nitrogen mineralized under anaerobic and aerobic conditions for some agricultural and forest soils of Washington. *Comm. Soil Sci. Plant. Anal.* 12:997–1009.
Sollins, P., G. Sphycher, and C.A. Glassman. 1984. Net nitrogen mineralization from light- and heavy-fraction forest soil organic matter. *Soil Biol. Biochem.* 16:31–37.
Sparling, G.P., and D.J. Ross. 1988. Microbial contributions to increased nitrogen mineralization after air-drying of soils. *Plant and Soil* 105:163–167.
Stanford, G., and W. Demar. 1969. Extraction of soil organic nitrogen by autoclaving in water: I. The NaOH-distillable fraction as an index of nitrogen availability in soils. *Soil Sci.* 107:203–205.
Stanford, G., and E. Epstein. 1974. Nitrogen mineralization-water relations in soils. *Soil Sci. Soc. Am. Proc.* 38:103–107.
Stanford, G., and S. Smith. 1972. Nitrogen mineralization potentials of soils. *Soil Sci. Soc. Am. Proc.* 36:465–472.
Stanford, G., J. Carter, and S. Smith. 1974. Estimates of potentially mineralizable soil nitrogen based on short-term incubations. *Soil Sci. Soc. Am. Proc.* 38:99–103.
Strader, R., D. Binkley, and C. Wells. 1988. Nitrogen mineralization in high elevation forests of the Appalachians. I. Regional patterns in southern spruce-fir forests. *Biogeochem.* in press.
Talpaz, H., P. Fine, and B. Bar-Yosef. 1981. On the estimation of N-mineralization parameters from incubation experiments. *Soil Sci. Soc. Am. J.* 45:993–996.
Thorne, J., and S. Hamburg. 1985. Nitrification potentials of an old-field chronosequence in Campton, New Hampshire. *Ecology* 66:1333–1338.
Thorne, J., A. Friedland, E. Miller, and J. Battles. 1987. Nitrification and nitrogen mineralization in an Adirondack spruce-fir sere. *Bull. Ecol. Soc. Am.* 68:429.

Turner, D.P., and E.H. Franz. 1985. The influence of western hemlock and western redcedar on microbial numbers, nitrogen mineralization, and nitrification. *Plant and Soil* 88:259–267.

Van Cleve, K., and R. White. 1980. Forest-floor nitrogen dynamics in a 60-year-old paper birch ecosystem in interior Alaska. *Plant and Soil* 54:359–381.

Van Praag, H., and F. Weissen. 1973. Elements of a functional definition of oligotroph humus based on the nitrogen nutrition of forest stands. *J. Appl. Ecol.* 10:569–583.

Van Schreven, D.A. 1967. The effect of intermittent drying and wetting of a calcareous soil on carbon and nitrogen mineralization. *Plant and Soil* 26:14–32.

Virzo De Santo, A., Al Alfani, and A. Fioretto. 1982. Nitrogen mineralization in southern beech forests. *Pedobiologica* 23:348–357.

Vitousek, P. 1982. Nutrient cycling and nutrient use efficiency. *Am. Naturalist* 119:553–572.

Vitousek, P.M., and S.W. Andraiese. 1986. Microbial transformations of labelled nitrogen in a clear-cut pine plantation. *Oecologia* (Berl.) 68:601–605.

Vitousek, P., and P. Matson. 1985. Disturbance, nitrogen availability, and nitrogen losses in an intensively managed loblolly pine plantation. *Ecology* 66:1360–1376.

Vitousek, P.M., J.R. Gosz, C.C. Grier, J.M. Melillo, and W.A. Reiners. 1982. A comparative analysis of potential nitrification and nitrate mobility in forest ecosystems. *Ecol. Monogr.* 52:155–177.

Vogt, K., C. Grier, and D. Vogt. 1986. Production, turnover, and nutrient dynamics of above- and belowground detritus of world forests. *Advan. Ecol. Res.* 14:303–337.

Voroney, R.P., and E.A. Paul. 1984. Determination of k_C and k_N *in situ* for calibration of the chloroform fumigation-incubation method. *Soil Biol. Biochem.* 16:9–14.

Waring, R.H., K. Cromack, Jr., P. Matson, R. Boone, and S. Stafford. 1987. Responses to pathogen-induced disturbance: Decomposition, nutrient availability, and tree vigour. *Forestry* 60:219–227.

Waring, S.A., and J.M. Bremner. 1964. Ammonium production in soil under waterlogged conditions as an index of nitrogen availability. *Nature* 201:951–952.

Weber, M.G., and K. Van Cleve. 1984. Nitrogen transformations in feather moss and forest floor layers of interior Alaska black spruce ecosystems. *Can. J. For. Res.* 14:278–290.

Webster, R. 1985. Quantitative spatial analysis of soil in the field. *Advan. Soil Sci.* 3:1–70.

Webster, S. 1984. Comparison of available nitrogen procedures for Douglas-fir (*Pseudotsuga menziesii*) soils. In: *Nitrogen assessment workshop*, May 19–20, 1982, RFNRP Report No. 2, College of Forest Resources, Univ. of Washington, Seattle, p. 41.

Youngberg, C. 1978. Nitrogen mineralization and uptake from Douglas-fir forest floors. *Soil Sci. Soc. Am. J.* 42:499–502.

Zöttl, H. 1960. Dynamik der Stickstoffmineralisation im organishen Waldbodenmaterial. II. Einfluss des stickstoffgehaltes auf die Mineralstickstoff-nachlieferung. *Plant and Soil* 13:183–206.

Carbon Sources for Bacterial Denitrification

E.G. Beauchamp* J.T. Trevors†, and J.W. Paul*

I. Introduction

There are many factors that affect denitrification and these have been reviewed by Delwiche and Bryan (1976), Payne (1981), and Knowles (1982). None of these authors have provided an in-depth review of the critical role of organic carbon in the denitrification process. Organic carbon availability is one of the most important factors that affects denitrifying activity in soil, yet the chemistry

*Present address: Department of Land Resource Science, Ontario Agricultural College, University of Guelph, Guelph, Ontario N1G 2W1
†Present address: Department of Environmental Biology, Ontario Agricultural College, University of Guelph, Guelph, Ontario N1G 2W1

Advances in Soil Science, Volume 10

of soil organic matter is only partially understood and its complexity has probably inhibited studies. A readily available C substrate is often added to denitrifying systems in the laboratory to ensure an electron supply when other variables are studied.

The production of gaseous nitrogen (N_2) by the reduction of nitrogenous oxides by the respective reductases in bacteria is known as biological denitrification:

$$\underset{\text{(nitrate)}}{NO_3^-} \rightarrow \underset{\text{(nitrite)}}{NO_2^-} \rightarrow \underset{\text{(nitric oxide)}}{NO} \rightarrow \underset{\text{(nitrous oxide)}}{N_2O} \rightarrow \underset{\text{(dinitrogen)}}{N_2}$$

Some denitrifying bacterial strains possess the complete pathway for denitrification, whereas others may only be capable of catalyzing one or several steps of the pathway (Ingraham, 1981). However, by partially contributing to the pathway, these organisms still contribute to the overall process of bacterial denitrification. Ingraham (1981) summarized the reasons for partial denitrification as follows: (1) metabolic intermediates (NO_2^-, NO, N_2O) of the pathway are available, but the initial substrate NO_3^- is absent; (2) environmental conditions such as excess O_2, unfavorable pH, or an elevated concentration of one pathway intermediate could inhibit one or more reactions; (3) differences in the induction rates of the reductase enzymes; and (4) genetically deficient bacterial strains not capable of synthesizing the complete complement of necessary reductases. For example, certain bacterial strains may only be capable of reducing NO_3^- to NO_2^-, NO_3^- to N_2O, NO_2^- through to N_2, or reduce NO_3^- to NO_2^- and NO to N_2O.

Most denitrifying bacteria are aerobic heterotrophs that belong to taxonomically diverse genera like *Pseudomonas*, *Alcaligenes*, and *Azospirillum*. In addition, some autotrophic organisms such as *Paracoccus denitrificans*, *Thiobacillus denitrificans*, and *Rhodopseudomonas sphaeroides* are also capable of denitrification (Table 1). The rates of denitrification in soil are controlled by numerous biological, chemical, and physical factors that are constantly changing both in time and space (Table 2). For example, the number and activity of denitrifying bacteria are controlled by the level of organic matter, oxygen concentration, gas diffusion rates, pathways of diffusion, moisture, temperature, pH, and substrate availability (nitrate and nitrite) (D.G. Evans et al., 1985; Focht and Chang, 1975; Nakajima et al., 1984a, 1984b; Payne 1981; C.J. Smith and Patrick, 1983; Tiedje, 1982; Trevors, 1985; Trevors and Starodub, 1988). As most denitrifying bacteria are aerobic heterotrophs, O_2 concentrations and carbon availability (see Table 3 for a summary of C sources found in soil) are extremely important parameters that influence microbial denitrification.

The complexity of substrate C supply in the soil is enhanced by the range of microorganisms that are capable of denitrification in soils. Knowles (1982), in his review of denitrifying bacteria, noted that species from the genera *Alcaligenes* and *Pseudomonas* are ubiquitous (Gamble et al., 1977). The specific

Table 1. Known denitrifying bacteria

Organism	Reference
Agrobacterium	Payne, 1981
Alcaligenes eutropha	Pfitzner and Schlegel, 1973
Azospirillum	Payne, 1981
Bacillus	Payne, 1981
Chromobacterium	Payne, 1981
Cytophaga sp.[a]	Adkins and Knowles, 1984
Flavobacterium	Payne, 1981
Hyphomicrobium	Payne, 1981
Neisseria	Payne, 1981
Paracoccus denitrificans	Verhoeven *et al.*, 1954; Bovell, 1967; Calder *et al.*, 1980
Propionibacterium	Payne, 1981
Pseudomonas denitrificans ATCC 13867	Nishimura *et al.*, 1980
Pseudomonas perfectomarinus	Balderston *et al.*, 1976
Pseudomonas sp. G59	Aida *et al.*, 1986
Pseudomonas spp.	Ishaque *et al.*, 1973; Garcia *et al.*, 1977; Auling *et al.*, 1978
Rhizobium	Payne, 1981
Rhodopseudomonas	Payne, 1981
Spirillum	Payne, 1981
Thiobacillus denitrificans	Ishaque and Aleem, 1973; Aminuddin and Nicholas, 1974a, 1974b
Thiomicrospira denitrificans	Timmer-Ten Hoor 1975, 1976, 1977
Unidentified soil isolate[a]	Trevors and Beauchamp, 1985
Vibrio	Payne, 1981

[a] Organisms capable of reducing N_2O to N_2 in the presence of sulfide and concentrations of acetylene normally inhibitory to N_2O reduction.

substrate C requirements for these species has not been documented. It may be expected that wide ecological diversity exists not only between species but also between strains within species. Although this does not provide much encouragement for definitive studies of denitrification in soils, progress may be made only when we are able to predict substrate C availability in relation to the denitrifying potential of the soil.

Because a large number of denitrifying bacteria are heterotrophs, it is not surprising that the nature and availability of carbon sources are significant factors that influence denitrification. In many pure culture investigations, readily metabolized compounds such as glucose (Payne, 1981) and formate (Nishimura et al., 1980) are used as sources of carbon. In addition, the sources and concentrations of C relative to NO_3^- and O_2 levels also control the rates of denitrification. Oxygen has an immediate and reversible effect on NO_3^- respiration by denitrifying cultures of *Pseudomonas aeruginosa* (Hernandez and Rowe, 1987). Oxygen inhibited NO_3^- utilization by preventing its uptake. However, NO_3^-

Table 2. Factors that influence the denitrification process other than C substrate supply

Factor	Organism or medium	Reference
Bacterial populations		
Generally accepted that denitrifiers are present in all soils. However, accurate estimates of denitrifying bacteria in soil are still difficult to obtain.		
Carbon sources		
Glucose had no influence on N_2O reduction	Soil	D.G. Evans *et al.*, 1985
CO_2 stimulated N_2O reduction	Soil	D.G. Evans *et al.*, 1985
Organic matter buffers chemical release of NO_2^-	Soil	Nelson and Bremner, 1969
Eh (electropotential)		
N content decreased as Eh dropped to −100 mV	Soil	Patrick, 1960
Temperature and O_2		
Synergistic effect of temperature and O_2 on denitrification	Soil	Misra *et al.*, 1974
Moisture		
N_2O evolution in soil was greater under fluctuating moisture conditions	Soil	C.J. Smith and Patrick, 1983
The greater the H_2O content, more rapid N_2O was converted to N_2	Sandy loam	Cady and Bartholomew, 1960
Rates of N loss directly related to H_2O content, especially near saturation	Soil	Pilot and Patrick, 1972; Rolston *et al.*, 1984
N_2O evolution was twofold higher in soil at 94% of its water-holding capacity (WHC) versus 51% WHC	Soil	M.S. Smith and Tiedje, 1979b
N amendments		
NO_2^- and NO_3^- amendments increased denitrification rates	Soil	Trevors, 1985
Nature of NO_3^- salt did not influence rate of NO_3^- release	Soil	Bremner and Shaw, 1958a
Oxygen		
Temporary low O_2 levels required for synthesis of nitrate and nitrous oxide reductases	*Pseudomonas*	Aida *et al.*, 1986

Table 2. (*Continued*)

Factor	Organism or medium	Reference
Oxygen (cont.)		
Phases of denitrification following O_2 depletion in soil	Soil	M.S. Smith and Tiedje, 1979b
NO evolution under anaerobic conditions was sixty-one-fold higher than in aerobically incubated soil	Loam soil	Johansson and Galbally, 1984
Denitrification decreased as O_2 increased; however, denitrification occurred under aerobic conditions	Oxidation ditch	Nakajima *et al.*, 1984a
Denitrification was not observed until O_2 concentration decreased to 0.21 μmol/ml	Soil	Trevors, 1985
Denitrification under aerobic incubation conditions	Sediment	Trevors and Starodub, 1988
pH		
Denitrification more rapid in pH 7–8 range.	Soil	Wijler and Delwiche, 1954; Nommik, 1956
More NO and N_2O are produced under acid conditions	Soil	Bollag *et al.*, 1973; Garcia, 1973
Denitrification occurred over pH range 3.5–11.2	Soil	Cady and Bartholomew, 1960; Davidson *et al.*, 1987
In NO_3-amended soil at pH 4.5 or less; N_2O evolution was instantaneous and nonbiological	Sandy loam	Cady and Bartholomew, 1960
NO_2^- decomposed to gaseous products in sterilized acidic soil	Soil	Nelson and Bremner, 1969, 1970
Soil amendments		
Liquid manure increased rate up to 23 g N/ha/day	Acid soil (pH 5.4)	Christensen, 1985
Sulfide		
Alleviation of acetylene blockage of N_2O reduction by sulfide	Soil	Evans *et al.*, 1985
Reduction of N_2O in the presence of sulfide and acetylene	Soil and soil bacterial isolate	Trevors and Beauchamp, 1985
Temperature		
Optimal denitrification occurred at 65°C	Soil	Nommik, 1956

Table 2. (*Continued*)

Factor	Organism or medium	Reference
Temperature (cont.)		
Rates of NO_3^- reduction were more affected by low temperatures than NO_2^- reduction	Soil	Bailey and Beauchamp, 1973
Denitrification minimal at 2 to 5°C, increasing with temperature up to 25°C. From 25 to 60°C actively increased slightly, then decreased significantly above 60°C	Clay and sandy soil	Bremner and Shaw, 1958a
Chemical denitrification occurs in frozen soil by soluble organic matter–NO_2^- reactions	Frozen soil	Christianson and Cho, 1983

reduction to NO_2^- in cell extracts was not inhibited by O_2 (Hernandez and Rowe, 1987). Furthermore, the inhibitory effect of O_2 was maximal at 0.2% O_2 saturation.

Because denitrifying bacteria can use NO_3^- and NO_2^- as terminal electron acceptors, specific energy-dependent transport systems should exist for these ions. Also, C sources may have a significant influence on the transport of NO_3^- and NO_2^- into the cells as electron flow to these compounds is coupled to adenosine triphosphate (ATP) formation (Gottschalk, 1986). Investigations of prokaryotic transport systems for NO_3^- and NO_2^- are limited (Brown et al., 1975; Kristjansson et al., 1978; Betlach et al., 1981). However, the uptake of NO_3^- appears to be an active transport process. There is also evidence suggesting that synthesis of the NO_3^- transport system proteins must be induced by growth in the presence of NO_3^-. Schloemer and Garrett (1974) and Betlach et al. (1981) observed that cells grown with NH_4^+ as the nitrogen source were unable to transport NO_3^-. Nitrite can function as a noncompetitive inhibitor of NO_3^- uptake, reducing the maximum velocity of uptake but not the half-saturation constant. However, Eppley and Coatsworth (1968) reported that NO_2^- did not inhibit NO_3^- transport in *Ditylum brightwelli*. The question of whether NO_2^- and NO_3^- are transported by the same system remains unclear.

Kristjansson et al. (1978) observed that the cytoplasmic pH value decreased in *P. denitrificans* and *Pseudomonas denitrificans* upon addition of NO_2^- to the culture medium. Possibly the NO_2^- entered the bacteria cells via the NO_2^-/H^+ symport or NO_2^-/OH^- antiport systems. In the presence of cyanide, transport was inhibited. The ATPase inhibitor, N,N^1-dicyclohexylcarbodiimide (DCCD), also inhibited NO_2^- transport about 3 minutes after its addition to cell suspensions. This suggests that ATPase was the proton transporter during NO_2^- uptake, as it is known that DCCD binds to the hydrophobic part of the ATPase enzyme and blocks the transmembrane proton movements catalyzed by the ATPase (Nalecz

Table 3. Some major C sources available for decomposition by bacteria in soil

Cellulose and its biodegradation product (glucose)
Chitin and its biodegradation product (N-acetylglucsamine)
Hemicellulose and its biodegradation products
arabinogalactans
arabinoxylans
arabinocuronoxylans
galactose
glucomannans
mannose
xylose
Inulin and its biodegradation product (fructose)
Lignin and its biodegradation products
conifer alcohol
coniferylaldehyde
ferulic acid
3-methoxy-4-hydroxyphenyl pyruvic acid
protocatechuic acid
quaicacylglycerol
vanillin
vanillic acid
Pectic substances
Pesticides
Proteins and amino acids
Simple sugars
Volatile fatty acids

et al., 1986). It appears that energy-dependent transport systems are required for NO_2^- and NO_3^- uptake. Therefore, carbon sources are a very important factor influencing both the uptake of NO_2^- and NO_3^- and the subsequent rates of denitrifying activity, as the formation of ATP from oxidative and substrate level phosphorylation requires metabolizable C-sources.

II. Carbon Sources

The following stoichiometric equation for denitrification is often cited with glucose as the C substrate.

$$5(CH_2O) + 4\,NO_3^- + 4H^+ \rightarrow 2N_2 + 5CO_2 + 7H_2O \qquad (1)$$

Accordingly, for each mole of CO_2 produced, about 0.8 mole NO_3^- is consumed. However, Payne (1981) suggested that a ratio of C (as glucose)/N (as

NO_3^-) of 3 or 4:1 was most effective in obtaining the denitrifying potential. On the other hand, Reddy et al. (1982) observed that the molar ratio of NO_3^- consumed/CO_2 produced ranged from 0.6 to 1.4 in several soils. They suggested that other electron acceptors may be involved or that CO_2 production may have occurred owing to fermentative processes. These discrepancies indicate the complexity of responses of denitrifying organisms to C substrates.

Denitrifying bacteria use a wide variety of organic acids, carbohydrates, and other organic compounds as carbon and energy sources when growing under aerobic conditions. However, under denitrifying conditions, these organisms may be restricted to fewer carbon sources. For example, *Pseudomonas stutzeri* metabolizes cysteine, isoleucine, leucine, and valine when grown aerobically. However, under anaerobic conditions, the same amino acids are not used as carbon sources (Bryan, 1981). The transport systems for these amino acids may not function under anaerobic conditions, thereby limiting their entry into the cell.

The anaerobic metabolism of aromatic substances under denitrifying conditions has also been reported by W.C. Evans (1977), Sleat and Robinson (1984) and Young (1984) (see Table 4). A number of studies have investigated the biodegradation of benzoate and its intermediates (Taylor et al., 1970; R.J. Williams and Evans, 1975; Aftring and Taylor, 1981). In addition, the anaerobic metabolism of halogens (Schennen et al., 1985) and substituted benzoates (Taylor, 1983) have been studied under anaerobic conditions. In many of these studies, pseudomonads are often responsible for the metabolism of the organic compounds. It is noteworthy that pseudomonads are also capable of denitrification when the O_2 concentration is reduced to some low critical value. This type of enzymatic versatility allows these organisms to metabolize a wide variety of C sources in the absence of oxygen.

The electron supply per mole of C from different substrates may be a factor determining their efficiency in denitrification. Carbon dioxide production from glycerol was about one-half of that from glucose, mannitol, and sucrose, although denitrification rates were similar (deCatanzaro and Beauchamp, 1985). This may indicate that the electron supply per mole of C in glycerol was greater than from the other sources. Assuming that electron supply is related to H^+ ion incidence, the C/H ratio in glycerol is about 25% less than that in glucose. Jacobson and Alexander (1980) found both succinate and methanol to be less stimulatory than glucose as C substrates for soil denitrifiers. Although the C/H relationship may hold for glucose and succinate, it did not hold for methanol. Other factors such as substrate acclimatization need to be taken into account.

Hansen and Kallio (1957) showed that molecular oxygen must be present for hydrocarbon utilization by *P. stutzeri* under anaerobic conditions. Denitrification did not occur with *n*-dodecane or 1-dodecene but occurred with dodecanol, dodecanal, and dodecanoic acid as C substrates. All of these compounds are readily metabolized under aerobic conditions. Whether hydrocarbons can be cooxidized under anaerobic conditions with other C substrate requires further investigation (Perry, 1979).

Table 4. Some selected examples of aromatic compounds metabolized under anaerobic conditions

Compound	Organisms	References
Phthalic acids	*Bacillus* sp. mixed culture	Afring and Taylor, 1981; Afring *et al.*, 1981
p-Cresol	Bacterial coculture (unidentified)	Bossert *et al.*, 1986
Benzoate	*Moxaxella* sp.	W.C. Evans, 1977, R.J. Williams and Evans, 1975
	Pseudomonas aeruginosa	Balba and Evans, 1980
	Desulfovibrio vulgaris	
Phenol	Bacterial consortium	W.C. Evans, 1977
Phloroglucinol	*Rhodospeudomonas gelatinosa*	W.C. Evans, 1977
Trihydroxybenzenes	*Pelobacter acidigallici*	Schink and Pfennig, 1981
Anaerobic degraduation of aromatic compounds	Denitrifying bacterial strain	Taylor and Heeb, 1972
Monofluorobenzoates and monochloro benzoates	Denitrifying bacterial strain	Taylor *et al.*, 1979
Lignin and lignin-derived compounds	Various anaerobes	Young and Frazer, 1987

A notable observation was made by Nishimura et al. (1980) with *P. denitrificans* ATCC 13867 supplied with formate. The growth of this organism was significantly stimulated with this C source along with a marked depression in NO_2^- accumulation. This was attributed to a lower nitrate reductase activity and higher nitrite reductase activity in the presence of formate. This alteration in enzyme activity was suggested to prevent NO_2^- toxicity and thereby enhance cell growth.

Bremner and Shaw (1958b) observed that glucose, sucrose, mannitol, and citrate were equally and highly effective in promoting denitrification. They concluded that the availability of C substrates was dependent on both their decomposability and their ability to support microbial growth. It is interesting to note, however, that cellulose eventually provided a good C substrate although the denitrification rate was initially very slow.

M.S. Smith and Tiedje (1979a) added either glucose or succinate to a soil in which corn had been planted. They observed that the denitrifying activity was much greater with glucose than succinate, especially in the soil in which corn had been grown. In fact, succinate reduced denitrification in this soil but not in the soil in which plants were not grown. They did not offer any explanation for this observation.

Yeomans and Bremner (1986) reported that ethanol was used more effectively than glucose or mannitol by denitrifiers. They warned that solvents such as acetone, methanol, or ethanol are difficult to remove from soil and may solubilize a fraction of the soil C, making it available to denitrifiers.

Marshall (1979) studied denitrification with several C sources with a miscible displacement apparatus. He observed the rate of denitrification to decrease as the degree of unsaturation of three 6-carbon, straight-chain alcohols increased. These compounds included 1-hexanol, *trans*-2-hexene-1-ol, and 2,4-hexadiene -1-ol. A similar trend was observed for cyclic hexanols (cyclohexanol and phenol). He observed that microbial activity also decreased as the degree of unsaturation increased. Rates with the ring hexanols were lower than with the straight-chain hexanols, presumably owing to a need for ring cleavage.

Compounds that may be available for decomposition in soils are shown in Table 3. Relatively few of these have been studied as C substrates for denitrifiers under laboratory conditions. Their wide diversity suggests that much remains to be researched if we are to obtain an understanding of C substrate availability for denitrifiers in soils. In addition, if the diversity of metabolic (catabolic) pathways are taken into account, it is evident that much remains to be known about the efficiency of various C substrates as electron donors for denitrifiers. The transfer of electrons from different C substrates through metabolic pathways under low O_2 and anaerobic conditions has not been well elucidated. It is evident that although the efficiency of different simple compounds as sources of electrons has not been extensively studied, there are unexplainable complexities that remain to be researched.

Several researchers have studied C as a factor that affects the denitrification process. Craswell (1978) warned that NO_3^- disappearance may be partly related to immobilization and not necessarily to denitrification, especially if N-free C sources (e.g., glucose) are added to the soil. Also Guenzi et al. (1978) observed increased denitrification in soil when treated with cattle manure. They attributed this to the creation of anaerobic microsites capable of promoting denitrification. In a field study, Aulakh et al. (1984) found that incorporation of straw in a conventionally tilled soil or application of straw to the surface of a zero-tilled soil doubled gaseous N losses. In addition to providing a C substrate source, the straw on the surface increased the H_2O content of the soil, thereby reducing O_2 supply. These few studies illustrate the complexity of denitrifier response to C sources under field conditions.

III. Natural C Substrates

Plant tissues, manure, and soil organic matter, or humus, have often been studied as sources of C for denitrifiers. The response of denitrifiers to these sources is expected to be complex, mainly owing to the fact that decomposition products and populations of microorganisms involved are not well understood. Further, the kind of decomposition products and their rate of production will vary depending on the O_2 status of the soil. The decomposition of complex C sources under anaerobic conditions was studied many years ago by Tenney and Waksman (1930). They concluded that decomposition under anaerobic soil conditions was much slower than under aerobic conditions and that lignins and organic N complexes were much more slowly decomposed than celluloses or hemicelluloses. It follows that decomposition and hence C substrate availability of natural materials will depend on the lignin content.

IV. Coculture and Multiculture Systems

Rhee and Fuhs (1978) studied a wastewater laboratory setup in which they isolated two organisms capable of denitrifying methanol. *Vibrio extorquens* was capable of excreting approximately one-half of methanol consumed as soluble metabolites, mostly citrate and isocitrate. A denitrifying organism, *P. stutzeri*, was then able to use the excreted metabolites. This is an example of a symbiotic relationship in which a one-carbon source was metabolized to supply C substrate to an organism requiring a more complex C substrate.

Another example of interspecies dependence was reported by Bossert et al. (1986). They isolated from polluted river sediment two bacterial species that were capable of using *p*-cresol as their sole carbon source when cocultured under NO_3^--reducing conditions. Nitrate served as the terminal electron acceptor for both organisms when incubated anaerobically. Nitrate was reduced to NO_2^-, then N_2O and N_2. One strain, PC-07, degraded the *p*-cresol to *p*-hydroxybenzoate, which was further degraded by the second organism, PB-04 (Figure 1). However,

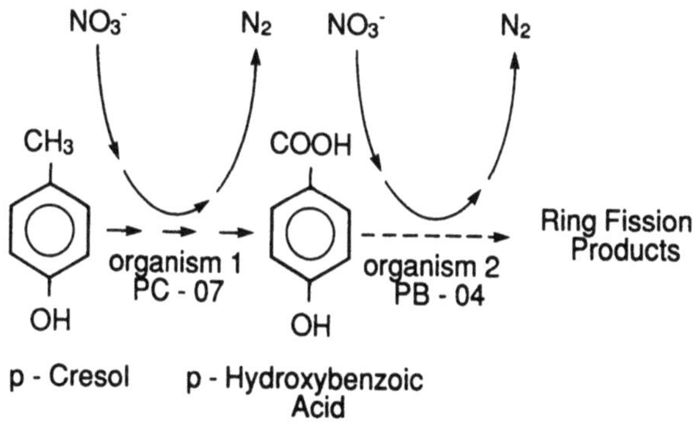

Figure 1. A schematic representation of anaerobic *p*-Cresol metabolism by a denitrifying bacterial coculture. (Reprinted with permission from *FEMS Microbiology Ecology*, vol. 38. I.D. Bossert, M.D. Rivera, and L.Y. Young, p-Cresol biodegradation under denitrifying conditions: isolation of a bacterial coculture. 1986. Elsevier Science Publishing Company, Inc.)

PB-04 was not capable of metabolizing *p*-cresol. Neither *p*-cresol nor *p*-hydroxybenzoate were degraded anaerobically when NO_3^- was absent. These examples illustrate a complexity in xenobiotic C-substrate utilization by soil denitrifiers. The question remains as to the extent of such incidences of mixed culture interactions under natural conditions. It also raises the question of one of the functions of NO_3^- in the soil environment. It may be necessary to consider NO_3^- leaching to groundwater desirable, for example, where it may be used by denitrifiers that degrade undesirable organic molecules.

V. Decomposition of Organic Materials and Denitrification

A. Animal Manures

Deherain (1897) observed greater disappearance of nitrate in soil amended with straw, fresh manure, and horse manure. Ferguson and Fred (1909) reported increased denitrification losses in manured soils. Many other researchers have, at least in part, attributed significant quantities of unrecovered nitrogen in manured soils to denitrification losses (Olsen et al., 1970; Wallingford et al., 1975; Guenzi et al., 1978; J.B. Cooper et al., 1984). Nitrous oxide, an intermediate in the denitrification process, was shown to increase in manured soils (Burford et al., 1976; Christensen, 1985; Egginton and Smith, 1986b).

Addition of animal wastes to soil may promote denitrification by creating anaerobic environments as well as providing an energy source to denitrifying bacteria. Anaerobic environments result from decreased oxygen supply to the soil atmosphere and increased consumption of oxygen by soil microorganisms and plant roots. The gas diffusion rates may be reduced in soil by increased soil moisture contents that result from slurry application (Comfort et al., 1986) and the blocking of soil pores by the fine solids fraction of manure (Stevens and Cornforth, 1974; Burford et al., 1976).

The application of manure increases the soluble C content of the soil (Meek et al., 1974; Comfort et al., 1986). Meek et al. (1974) reported very little NO_3^- present below 80 cm in an irrigated desert soil, even though very high levels of NO_3^- were produced following heavy applications of manure. They found increased quantities of soluble C as far as 140 cm below the surface. They concluded that this water-soluble C increased denitrification under reducing conditions at lower depths in the soil profile. Elliot and McCalla (1972) reported high concentrations of methane down to the 152-cm depth in soils beneath a feedlot, indicating that C sources for methane production percolated to these depths. They suggested that the reducing conditions and soluble C at these depths were the cause of low concentrations of NO_3^- in the shallow water table below the feedlot.

Sherwood (1980) reported pig slurry to be a readily available C source for denitrifiers. She compared nitrate disappearance in soil amended with pig slurry and nitrate to soil amended with glucose and nitrate. The NO_3^- in the slurry-treated soil disappeared almost as rapidly as in the glucose-treated soil.

Manure contains mainly the undigested remains of the animal feed, with a high microbial population. We would not expect compounds that have resisted decomposition in the animal and during storage to be readily available C sources for denitrifying bacteria. The change in microbial environment after manure application to the soil may change the population dynamics, killing off some of the microorganisms, and thus making microbial C available to denitrifying bacteria.

When manure is stored anaerobically as a slurry, the undigested plant residues undergo further fermentation, resulting in the accumulation of fermentation products. Products of anaerobic fermentation of manure are mainly volatile fatty acids. The volatile fatty acids reduced are mainly acetic, propionic, and butyric acids, with smaller concentrations of isobutyric, valeric, and isovaleric acids. Significant concentrations of these acids have been reported in swine slurry (Yashura and Fuwa, 1977; P. Cooper and Cornforth, 1978; A.G. Williams, 1983; Rainville and Morin, 1985) and in cattle slurry (Schuman and McCalla, 1976; P. Cooper and Cornforth, 1978; Guenzi and Beard, 1981; Patni and Jui, 1985).

Could these acids serve as C sources for denitrifying bacteria in soil? A.G. Williams (1983) reported a significant correlation ($r = 0.94$) between volatile organic acids in pig slurry and the biochemical oxygen demand of the slurry. Abufayed and Schroeder (1986) reported a significant correlation between acetate disappearance and NO_3^- and NO_2^- disappearance when acetate was added to sewage sludge as a C source. This suggests that volatile fatty acids when present in manure may be readily available C sources for denitrifiers in soil.

The pathways of acetate and propionate degradation have been reported for *Escherichia coli* in aerobic environments (Wegener et al., 1968), but they have not been studied in microorganisms by using nitrate as a terminal electron acceptor.

B. Plant Residues

The addition of plant residues to soil under anaerobic conditions results in increased N losses by denitrification. Increased rates of denitrification were observed with alfalfa (*Medicago sativa*) amendments to soil (Jansson and Clark, 1952; Wijler and Delwiche, 1954; deCatanzaro and Beauchamp, 1985; Myrold and Tiedje, 1985) straw (Bremner and Shaw, 1958b; McGarity et al., 1958; Aulakh et al., 1984; deCatanzaro and Beauchamp, 1985), hay dust (Hutchinson, 1918), fresh crop residues (Ryden and Lund, 1980), and grass (Bremner and Shaw, 1958b). Although many researchers have reported increased denitrification losses in soil with crop residues, the dynamics of anaerobic decomposition with respect to denitrifying bacteria are not well understood.

Carbon compounds in plants are divided into five broad categories (Alexander 1977): (1) cellulose, 15 to 60% of the dry weight of the plant; (2) hemicellulose, 10 to 30%; (3) lignin, 5 to 30%; (4) water-soluble constituents, including simple sugars, amino acids, and aliphatic acids; (5) ether- and alcohol-soluble constituents, including fats, oils, waxes, and resins. Cellulose and hemicellulose compounds are the major components of most plants; therefore, their decomposition is important with respect to denitrifiers.

The literature contains evidence of interaction between denitrifying bacteria and fermentative cellulose decomposers during anaerobic decomposition of crop residues. During anaerobic decomposition of alfalfa, rye (*Secale cereale*) straw, and corn (*Zea mays* L.) stalks, the water-soluble fraction was most rapidly decomposed, followed by cellulose and hemicellulose (Tenney and Waksman, 1930). Lignin was only very slightly decomposed during the period studied. Harper and Lynch (1981) investigated anaerobic decomposition rates of oat (*Avena sativa*) straw in the field. The straw was composed of 7.3% water-soluble material, 35.5% hemicellulose, 42.1% cellulose, and 13.5% lignin. Within two months, almost one-half the cellulose and hemicellulose had disappeared. Anaerobic cellulase producers in soil are mainly bacteria of the genus *Clostridium* (Ljundahl and Eriksson, 1985). Cellulase-producing fungi, or actinomycetes, are not significant in anaerobic soils (Alexander, 1977).

The products of anaerobic decomposition of plant residues in soil are organic acids (Tenney and Waksman, 1930; Greenwood, 1961; Alexander, 1977). Organic acids are produced mainly by bacterial fermentation of the water-soluble fraction, cellulose, and hemicellulose according to Harper and Lynch (1981). They reported the dominant organic acid to be acetic acid. Mainly, butyric acid formed during decomposition of straw (Mishustin and Erofeev, 1964). Acetic and butyric acids were produced during corn residue decomposition (Moraghan and Ayotade, 1968). Formic, acetic, butyric, valeric, and lactic acids were produced during rice (*Oryza sativa*) straw decomposition (Gotoh and Onikura, 1971); acetic, propionic, and butyric from green manure (Chandrasekaran and Yoshida, 1973), wheat (*Triticum aestivum*) straw (Lynch, 1977a; Lynch and Gunn, 1978; Wallace and Elliot, 1979), and couch grass (*Agropyron repens*) rhizomes (Lynch et al., 1980). Acetic acid was measured during decomposition of barley (*Hordeum vulgare*), wheat, oat, and rape (*Brassica napus*) straw (Lynch, 1977b).

The formation of volatile fatty acids was reported to be reduced in the presence of NO_3^- (Greenwood and Lees, 1960). Greenwood (1961) suggested that the lack of organic acid accumulation may be due to NO_3^- being used preferentially as a hydrogen (electron) acceptor. Guenzi and Beard (1981) found that volatile fatty acids did not accumulate in cattle manure slurry until the redox potential fell to -100 mV. They suggested that if volatile fatty acids were formed at redox potentials above -100 mV, they were used by aerobic or facultatively aerobic respirers. Lynch and Gunn (1978) measured the accumulation of acetic acid during anaerobic decomposition of wheat straw. When NO_3^- was added to the anaerobic environment, the accumulated acetic acid rapidly disappeared, suggesting that the acetic acid provided a carbon source for denitrifying bacteria. The lack of accumulation of volatile fatty acids may be due to rapid oxidation in the presence of oxygen (Greenwood, 1961), nitrate (Lynch and Gunn, 1978), and sulphate (Laanbroek and Pfennig, 1981; Sorensen et al., 1891; Balba and Nedwell, 1982; Widdel and Pfennig, 1982). The literature is not clear whether organic acids continue to be produced in the presence of NO_3^- or whether they are oxidized by denitrifying bacteria as rapidly as they are produced.

Denitrifying bacteria may require anaerobic cellulose fermenters to decompose the cellulose and hemicellulose. This may have occurred when Bremner and Shaw (1958b) observed that cellulose eventually became a good source of C for soil denitrifiers. Payne (1981) includes an extensive summary of the denitrifying bacteria. Of these bacteria, only *Bacillus lichenformis* and *Pseudomonas fluorescens* var. *cellulosa* may produce cellulase (Ljungdahl and Eriksson, 1985). It is unlikely that these two species of bacteria are the dominant ones during crop residue decomposition in the presence of NO_3^-. We suggest a syntrophism of denitrifying bacteria with anaerobic cellulose fermenters in which the denitrifiers use the breakdown products of the cellulases as well as the products of fermentation. The utilization of the products of the cellulase enzymes prevent repression of the enzymes. The oxidation of the fermentation products, the organic acids and hydrogen, allows the fermentative bacteria to obtain more energy from the substrate.

Bell (1969) reported a drop of the redox potential to less than -300 mV within 2 days in a soil incubated anaerobically with glucose. The potential rose rapidly to $+200$ mV until 8 days of incubation. He reported that the low redox potential was probably produced by a redox equilibrium that involved H_2 produced by fermentative organisms. These observations were also made by Yamane and Sato (1964, 1968) in anaerobic soils amended with glucose. Onodera (1929) observed H_2 accumulation when green manure was added to a rice field. From these observations, Yamane and Sato (1964) concluded that H_2 was probably formed by *Clostridium* sp. and was usually observed when soluble carbohydrate or green manure was added to a rice field. When NO_3^- was added to these soils, the large drop in Eh was not observed (Yamane and Sato, 1968). They reported that the reduction of NO_3^- utilized H_2. The denitrifying bacteria, *P. denitrificans*, can use H_2 as an energy source (Payne, 1981).

The utilization of H_2 and organic acids by denitrifiers is advantageous to the fermentative bacteria because it provides more energy from the substrate. Mosey (1983) states that a carbohydrate fermenter would preferentially produce acetic acid and H_2 because it yields twice the energy as fermentation to propionic or butyric acids. The formation of propionic and butyric acids is the response of bacteria to accumulations of H_2 during surges of organic energy input, and the formation of butyric acid also reduces the acid load. This explains the different ratios of organic acids produced during anaerobic fermentation.

The observation of deCatanzaro and Beauchamp (1985) of rapid buildup of water-extractable C following nitrate disappearance during anaerobic decomposition of alfalfa and the continuous buildup of water-extractable C during anaerobic straw decomposition may be due to the accumulation of organic acids, the products of bacterial fermentation. Increased water-soluble C (WSC) was also observed during anaerobic decomposition of alfalfa in soil by Gale and Gilmour (1986). Organic acids accounted for 69% (Lynch, 1977a) and almost 100% (Lynch and Gunn, 1978) of the increased concentration water-soluble C compounds produced during anaerobic straw decomposition. Research is required to implicate the role of the major organic acids as C sources for denitrifiers, and the role of the denitrifier-fermenter interaction.

An important consideration with respect to the interaction of these two groups of bacteria is the estimation of C utilization for denitrification in a mixed bacterial culture system as found in soil. A more efficient C source may be one that is selectively used by denitrifying bacteria. An example is denitrification with acetate as a C source. It is a product of fermentation; therefore, it cannot be used as an energy source for fermentative bacteria. Glucose can be used by fermentative bacteria and denitrifiers alike. McCarty et al. (1969) observed a more efficient NO_3^- removal from wastewater with acetate than with glucose, even though three times as much C is required to produce microbial biomass from acetate as with glucose (Stouthamer, 1976).

VI. Denitrification in Relation to Measured Available Soil C Substrate

Burford and Bremner (1975) found that although denitrification capacity was significantly correlated with total organic C in a range of soils, a very high correlation was obtained with water-soluble organic C or mineralizable C (Figure 2). They concluded that denitrification in soils under anaerobic conditions is controlled largely by the supply of readily decomposable organic matter. They also showed a very high correlation between mineralizable C and WSC. The amount of mineralizable C agreed stoichiometrically with the amount of NO_3^- denitrified, but the quantity of WSC present was insufficient to carry out denitrification to the extent measured. In fact, under their incubation conditions, 28% had to come from the water-insoluble forms of organic carbon.

Beauchamp et al. (1980), in a study involving the A, B, and C horizons of several soils of different textures and drainage classes, found the greatest correlation with total organic C. Water-extractable C, mineralizable C (CO_2 production), and 0.05 *M* $Ba(OH)_2$-extractable C were unsatisfactory for the B and C horizons. All methods for estimating available C substrate were satisfactory for soils of well-drained or imperfectly drained classes, but only total organic C and 0.05 *M* $Ba(OH)_2$-extractable C were satisfactory for poorly drained soils. It appeared that, with water-extractable C, there were two relationships: one involving soils that were mostly poorly drained; another that included mostly well-drained or imperfectly drained soils. It thus appears that some attention needs to be paid to the drainage status of the soils if extractable C is to be used as a measure of available C substrate. With both of the relationships, extrapolation of denitrification rate versus water-extractable C relationship suggested that a significant fraction of the water-extractable C (40 to 100 μg C/g soil) appears to be unavailable to denitrifiers. Similarly, Reddy et al. (1982) observed that approximately 25% of the water-extractable C was available apparently at a slower rate. This has also been observed by Burton and Beauchamp (1985) who

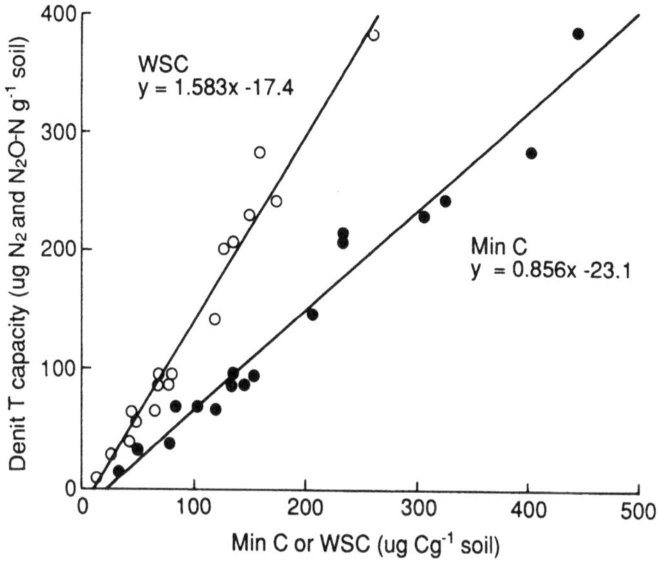

Figure 2. Denitrifying capacity in relation to water-soluble carbon (WSC) and mineralizable carbon (Min C) in 17 soils. (Reprinted with permission from *Soil Biology and Biochemistry*, vol. 7. J.R. Burford and J.M. Bremner, Relationships between denitrification capacities of soils and total, water-soluble and readily decomposable soil organic matter. 1975. Pergamon Journals, Ltd.

noted that water-extractable C in excess of 60 to 80 μg C/g OD soil was required to sustain measurable denitrifier activity using the acetylene block technique. Nothing is known about the composition of water-extractable C that may explain these observations. Davidson et al. (1987) used the denitrification potential of different acid forest soils to evaluate methods for estimating the availability of soil C to microorganisms. They concluded that C mineralization under aerobic conditions provided the best measure of available C. They included cold water-soluble C, boiling water-extractable C, and total C in their study.

Germon et al. (1981) studied 20 soils in which they found that drying significantly increased the rate of denitrification. They also found biomass C measurements (Jenkinson and Powlson, 1976) were more highly correlated with the rate of denitrification—especially if the soils were not air-dried previously—compared with WSC (20 and 100 C or autoclaved at 120 C), which, in turn, was a better indicator of available C than total organic C. Germon et al. (1981) explained either that biomass C became readily available following soil saturation and supported denitrification or that the availability of C to the denitrifiers was proportional to the active biomass mineralizing the organic matter. Myrold and Tiedje (1985) observed that microbial biomass—measured by fumigation or microbial ATP extraction—and denitrification capacity increased together when ground alfalfa tissues were added as a C source. Further, denitrifiers appeared to be able to compete successfully for the C source. The measurement of available

C substrate for denitrifiers in soil has been variably successful in some studies. It would not be expected that such rapid methods would be successful in soils to which crop residues or manures were recently added, for reasons suggested in Section V. Patten et al. (1980) showed that air drying and air-dry storage of soils could greatly increase the rate of denitrification. They suggested that this effect was due to increased C substrate availability. Similar observations with other processes involving the microbial biomass indicate a need to take soil preparation into account in laboratory studies.

VII. Denitrification Near Roots

The contributions of plants or plant roots to denitrification has been studied. Stefanson (1972) suggested that plants increase O_2 demand and also supply easily decomposable organic matter, both of which would increase the denitrification potential. Also, Woldendorp (1963) and Bailey (1976) found that plant roots stimulated loss of soil N by denitrification. However, Mosier (1986) reported that under soil water conditions usually encountered during plant growth, insufficient soluble C was derived from actively growing plants to sustain denitrification. Similarly, Haider et al. (1985) found that corn (*Z. mays* L.) or wheat (*T. aestivum*) plants did not stimulate denitrification although 60% of root biomass C was rhizodeposited by the termination of experiments. It was thought that sufficient C was exuded by the plant roots to promote denitrification. Later, Haider et al. (1987) confirmed these observations although additional NO_3^- was added to ensure an adequate supply to actively growing corn plants. They observed, however, that denitrification occurred in the later stages of corn growth as the roots began to decompose. These studies do not clearly define the conditions near plant roots that would be conducive to denitrifier activity. For example, although C and NO_3^- supplies may be adequate, the O_2 supply may also be high despite root respiration. The plant roots would absorb H_2O from the soil pores, opening them for enhanced gas diffusion to and from the rhizosphere regions. M.S. Smith and Tiedje (1979a) showed that the potential denitrifying activity decreases rapidly in the first few millimeters away from roots; this was attributed to the availability of organic matter. Their research also suggested that enhancement of denitrification near plant roots occurred only at relatively high NO_3^- levels.

VIII. Acetylene as a C Substrate for Denitrifiers

The use of acetylene as an inhibitor of N_2O reduction is considered a significant breakthrough in easing the measurement of the denitrification process (Balderston et al., 1976; Yoshinari and Knowles, 1976). The possibility of acetylene as a possible C source for denitrifiers was first reported by Germon (1980). He noted that NO_3^- disappearance under anaerobic soil conditions was greater in the presence of acetylene than in its absence. This observation was substantiated later by Yeomans and Beauchamp (1982). Meanwhile Watanabe and deGuzman (1980) showed in their studies of denitrification that acetylene disappeared in paddy soils during prolonged anaerobic conditions. They noted that NO_3^- addition strongly inhibited acetylene disappearance. They also noted that ethylene formation was negligible and did not account for acetylene disappearance, but that formic, acetic, propionic, and butyric acids were present in acetylene-treated soil. Their work did not suggest that denitrifiers used acetylene directly as a C substrate source. As discussed earlier, denitrifiers may use low-molecular-weight fatty acids as C sources. It was subsequently shown by Gross et al. (1982) that commercial sources of acetylene may contain significant quantities of acetone that could serve as a C source for denitrifiers. Terry and Duxbury (1985) have demonstrated, using ^{14}C-labeled acetylene, that microorganisms in soils may easily adapt to acetylene metabolism after 7 days of aerobic or anaerobic incubation. Addition of ground-up alfalfa top tissues tended to slow microbial adaptation to acetylene in one soil. It thus appears that avoidance of long exposure to acetylene is recommended if denitrification and C sources are studied, providing that the acetylene does not contain C contaminants that would be available to denitrifiers. Recently, Egginton and Smith (1986a) found that acetylene increased CO_2 and decreased O_2 levels in the soil profile. They concluded that the effect was relatively small and probably would not affect denitrification although acetylene was used as a substrate.

IX. Denitrification Kinetics Involving C Substrates

Bowman and Focht (1974) assessed denitrification rates in two soils as a function of NO_3^- and glucose-C concentrations. In one soil (Coachella sand), rates of NO_3^- loss closely approximated Michaelis–Menton, or zero-order, kinetics, whereas in another soil (Hanford sandy loam) they did not. Failure to approximate Michaelis–Menton kinetics with the second soil was attributed to unsaturation of the enzyme systems in the denitrifying bacteria.

Considerable controversy arose in determining whether denitrification followed or could be described mathematically by zero- or first-order kinetics. For example, Stanford et al. (1975) observed that first-order kinetics described the disappearance of NO_3^- adequately, that is, the rate of decrease depended on the NO_3^- concentration. On the other hand, Smid and Beauchamp (1976) suggested that decreasing organic matter availability may also be responsible for this pattern of denitrification rate. They suggested that denitrification kinetics could be described by a zero-order equation when C was either readily available or seriously limiting, whereas a first-order equation was suitable where C availability changed from readily available to limiting during the study period. It would be expected probably that, under "natural" conditions, the supply of both NO_3^- and C substrates would influence the rate of denitrification. Reddy et al. (1982) reviewed the roles of NO_3^- and C supply on denitrification kinetics. They attempted to describe the rate of denitrification as a combined first-order process that involved both NO_3^- and available C concentrations. Available C was defined as either WSC or CO_2-C (maximum available carbon [Cmax]) released over a 14-day period during incubation of the soil in the dark at 30 C. They noted that WSC comprised 0.4 to 0.9% of the total organic C, whereas Cmax represented 0.6 to 1.4% of the total C in the soils. They showed that denitrification rates were proportional to the concentrations of NO_3^- and available C. They also concluded that the denitrification rate coefficient (*Kn*) was essentially independent of soil type among the nine mineral soils studied but possibly not for the organic soils included in the study. Their results generally showed that the rate of denitrification can be controlled by the availability of C. They concluded that under field conditions, available C is seldom expected to be a limiting factor to denitrify small amounts of NO_3^-.

More recently, Rolston et al. (1984) studied a mathematical model in which the denitrification rate was related to water-extractable C concentration as well as NO_3^- concentration, soil-water content approaching saturation, and temperature. They, too, used water-extractable organic C that included a component derived from the total organic C by decomposition (Reddy et al., 1980). They pointed out the lack of understanding of the contribution of C from manures and plant residues to denitrification. They found reasonable agreement between measured rates and total amounts of denitrification and those calculated with the model. It would seem that a model describing N and C transformations under anaerobic or alternate aerobic/anaerobic conditions is needed to describe denitrification in relation to C supply. Further elaboration of the NCSOIL model presented by Molina et al. (1983) may be useful in this respect.

X. Denitrification Versus Dissimilatory Nitrate Reduction

The availability of C may play a role in determining whether denitrification and/or dissimilatory nitrate reduction (DNR) to NH_4^+ occurs. Tiedje et al. (1982) postulated that the extent of DNR and denitrification depends on the *C*/*e*-acceptor ratio. With a high *C*/*e* ratio, DNR would likely predominate, whereas the opposite would occur with a low *C*/*e* ratio. This postulation is supported by Wijler and Delwiche (1954), who observed that NO_3^- reduction to NH_4^+ was only observed at extremely high substrate (alfalfa) levels under anaerobic conditions. This has been confirmed more recently by deCatanzaro et al. (1987), who also noted that sulfide could also induce DNR to become prevalent. They hypothesized that the creation of highly reduced conditions was conducive to DNR. In their study, they observed that alfalfa caused much higher levels of DNR than straw or glucose. The addition of Na_2S with glucose caused a considerable stimulation of the DNR process. It is noteworthy that Bloomfield (1969) observed a much higher reduction state in soil with alfalfa than rice leaves, as evidenced by the evolvement of free H_2S in the presence of added Na_2SO_4.

XI. Summary

It is generally acknowledged that organic C substrate supply in soils is one of the most important factors limiting the denitrification process. It is understandable that relatively little research has been done on this topic because of the complexity of organic matter and the large number of diverse microbial species involved. The objective of this chapter is to review available information and provide a guide for future research. Sources of C substrates in soil are summarized in Figure 3.

Many laboratory studies have been conducted that involve an artificial C substrate to supply denitrifiers with electrons. Relatively few studies have dealt with natural sources, that is, those one typically finds under natural conditions in the field. Perhaps the most common substrate used in the laboratory is glucose.

Those interested in denitrification have attempted simple measurements of available organic C substrate and correlated these values with denitrification rates mostly in the laboratory. In most denitrification studies, a relatively pure C substrate is often added to fulfill the need for rapid denitrification. This has not provided much insight into the relationship between C availability and denitrifier activity. In few cases have natural substances such as animal manures or crop residues been evaluated as C sources or, much less, studied in detail.

One of the most striking revelations of this review is the scarcity of information on the biodegradation of C compounds and natural substances in conjunction with denitrification under anaerobic conditions. Consequently, the role of fermentative organisms in providing C substrates for denitrifiers through biodegradation of complex C compounds, for example, remains to be elucidated. Several examples were found in the literature that demonstrated the reliance of a denitrifying species on another organism to provide a suitable substrate. The potential for denitrification in the field cannot be adequately estimated until C availability with crop residues and manures is better understood.

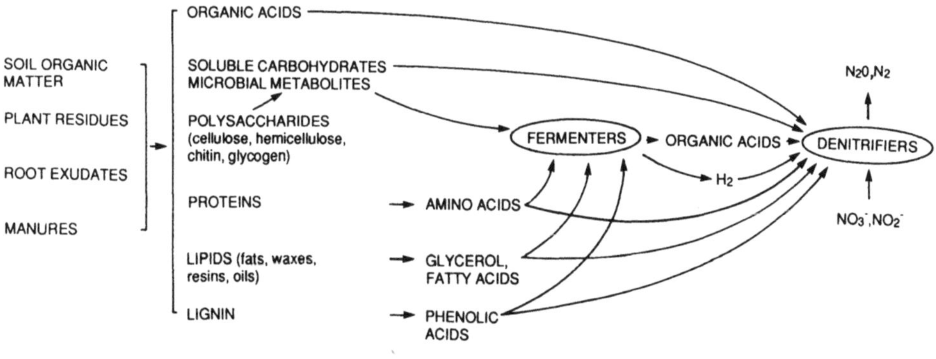

Figure 3. Schematic flow pathways of organic carbon substances to fermenters and denitrifiers under anaerobic conditions.

The application of zero-, first-, or higher-order models is limited by our ability to estimate organic C availability. Organic C supply plays an important role in dissimilatory NO_3^- reduction to NH_4^+ and its relationship with denitrification is intriguing. Denitrification near roots is a special case of local organic C supply. The changes in soil conditions induced by roots with water uptake and O_2 supply need to be considered in conjunction with C supply. Short-term water saturation of rooted soil should provide an interesting area for future denitrification research inasmuch as it is likely to be encountered periodically in the field.

Special consideration needs to be given to acetylene as a possible C source because it is widely used to measure denitrification. Although some researchers have reported its possible role as a C source, others have not noted this problem, especially in short-term studies.

Acknowledgments

This research was supported by the Natural Sciences and Engineering Research Council of Canada operating grants to E.B. and J.T.T. and by the Ontario Ministry of Agriculture and Food.

References

Abufayed, A.A., and E.D. Schroeder. 1986. Kinetics and stoichiometry of SBR/denitrification with a primary sludge carbon source. *J. Water Pollut. Control Fed.* 58:398–405.

Adkins, A.M., and R. Knowles. 1984. Reduction of nitrous oxide by a soil *Cytophaga* in the presence of acetylene and sulfide. *FEMS Microbiol. Lett.* 23:171–174.

Aftring, P.R., and B.F. Taylor. 1981. Aerobic and anaerobic catabolism of phthalic acid by a nitrate-respiring bacterium. *Arch. Microbiol.* 130:101–104.

Aftring, P.R., B.E. Chalker, and B.F. Taylor. 1981. Degradation of phthalic acids by a denitrifying mixed culture of bacteria. *Appl. Environ. Microbiol.* 41:1117–1183.

Aida, T., S. Hata, and H. Kusunoki. 1986. Temporary low oxygen conditions for the formation of nitrate reductase and nitrous oxide reductase by denitrifying *Pseudomonas* sp. G59. *Can. J. Microbiol.* 32:543–547.

Alexander, M. 1977. *Introduction to Soil Microbiology*, 2d ed., Wiley, NY. 467 pp.

Aminuddin, M., and D.J.D. Nicholas. 1974a. Electron transfer during sulphide and sulphite oxidation in *Thiobacillus denitrificans*. *J. Gen. Microbiol.* 82:115–123.

Aminuddin, M., and D.J.D. Nicholas. 1974b. Sulphide oxidation linked to the reduction of nitrate and nitrite in *Thiobacillus denitrificans*. *Biochim. Biophys. Acta* 325:82–93.

Aulakh, M.S., D.A. Rennie, and E.A. Paul. 1984. The influence of plant residues on denitrification rates in conventional and zero-tilled soils. *Soil Sci. Soc. Am. J.* 48: 790–794.

Auling, G., M. Reh, C.M. Lee, and H.G. Schlegel. 1978. *Pseudomonas pseudoflava*, a new species of hydrogen-oxidizing bacteria: Its differentiation from *Pseudomonas flava* and other yellow-pigmented, Gram-negative, hydrogen-oxidizing species. *Int. J. Syst. Bacteriol* 28:82–95.

Bailey, L.D. 1976. Effects of temperature and roots on denitrification in a soil. *Can. J. Soil Sci.* 56:79–87.

Bailey, L.D., and E.G. Beauchamp. 1973. Effects of temperature on NO_3^- and NO_2^- reduction, nitrogenous gs production, and redox potential in a saturated soil. *Can. J. Soil Sci.* 53:213–218.

Balba, M.T., and W.C. Evans. 1980. The anaerobic dissimilation of benzoate by *Psuedomonas aeruginosa* coupled with *Desulfobibrio vulgaris*, with sulphate as terminal electron acceptor. *Biochem. Soc. Trans.* 8:624–627.

Balba, M.T., and D.B. Nedwell. 1982. Microbial metabolism of acetate, propionate, and butyrate in anoxic sediment from the Colne Point saltmarsh, Essex, U.K. *J. Gen. Microbiol.* 128:1415–1422.

Balderston, W.L., B. Sherr, and W.J. Payne. 1976. Blockage by acetylene of nitrous oxide reduction in *Pseudomonas perfectomarinus*. *Appl. Environ. Microbiol.* 31:504–508.

Beauchamp, E.G., C. Gale, and J.C. Yeomans. 1980. Organic matter availability for denitrification in soils of different textures and drainage classes. *Comm. Soil Sci. Plant Anal.* 11:1221–1233.

Bell, R.G. 1969. Studies on the decomposition of organic matter in flooded soil. *Soil. Biol. Biochem.* 1:105–116.

Betlach, M.R., J.M. Tiedje, and R.B. Firestone. 1981. Assimilatory nitrate uptake in *Pseudomonas fluorescens* studied using nitrogen 13. *Arch. Microbiol.* 129:135–140.

Bloomfield, C. 1969. Sulphate reduction in waterlogged soils. *J. Soil Sci.* 20:207–221.

Bollag, J.M., S. Drzymala, and L.T. Kardos. 1973. Biological versus chemical nitrite decomposition in soil. *Soil Sci.* 116:44–50.

Bossert, I.D., M.D. Rivera, and L.Y. Young. 1986. p-Cresol biodegradation under denitrifying conditions: Isolation of a bacterial coculture. *FEMS Microbiol. Ecol.* 38:313–319.

Bovell, C. 1967. The effect of sodium nitrite on the growth of *Micrococcus denitrificans*. *Arch. Mikrobiol.* 59:13–19.

Bowman, R.A., and D.D. Focht. 1974. The influence of glucose and nitrate concentrations upon denitrification rates in a sandy soil. *Soil Biol. Biochem.* 6:297.

Bremner, J.M., and K. Shaw. 1958a. Denitrification in soil. I. Methods of investigation. *J. Agric. Sci.* 51:22–39.

Bremner, J.M., and K. Shaw. 1958b. Denitrification in soil. II. Factors affecting denitrification. *J. Agric. Sci.* 51:40–52.

Brown, C.M., D.S. MacDonald-Brown, and S.O. Stanley. 1975. Inorganic nitrogen metabolism in marine bacteria: Nitrate uptake and reduction in a marine pseudomonad. *Marine Biol.* 31:7–13.

Bryan, B.A. 1981. Physiology and biochemistry of denitrification. In: *Denitrification, nitrification and atmospheric nitrous oxide*, C.C. Delwiche, ed., Wiley, NY. pp. 67–84.

Burford, J.R., and J.M. Bremner. 1975. Relationships between denitrification capacities of soils and total, water-soluble and readily decomposable soil organic matter. *Soil Biol. Biochem.* 7:384–394.

Burford, J.R., D.J. Greenland, and B.F. Pain. 1976. Effects of heavy dressings of slurry and inorganic fertilizers applied to grassland on the composition of drainage waters and the soil atmosphere. In: "Agriculture and Water Quality," *Tech. Bull. Minist. Agric. Fisheries and Food*, No. 32, HMSO, London, pp. 432–443.

Burton, D.L., and E.G. Beauchamp. 1985. Denitrification rate relationships with soil parameters in the field. *Comm. Soil Sci. Plant Anal.* 16:539–549.

Cady, F.B., and W.V. Bartholomew. 1960. Sequential products of anaerobic denitrification in Norfolk soil. *Soil Sci. Soc. Am. Proc.* 24:477–482.

Calder, K., K.A. Burke, and J. LaScelles. 1980. Induction of nitrate reductase and membrane cytochromes in wild type and chlorate-resistant *Paracoccus denitrificans*. *Arch. Microbiol.* 126:149–153.

Chandrasekaran, S., and T. Yoshida. 1973. Effect of organic acid transformations in submerged soils on growth of the rice plant. *Soil Sci. Plant Nutr.* 19:39–45.

Christensen, S. 1985. Denitrification in an acid soil: effects of slurry and potassium nitrate on the evolution of nitrous oxide and on nitrate-reducing bacteria. *Soil. Biol. Biochem.* 17:757–764.

Christianson, C.B., and C.M. Cho. 1983. Chemical denitrification in frozen soils. *Soil Sci. Soc. Am. J.* 47:38–42.

Comfort, S.D., K.A. Kelling, D.R. Keeney, and J.C. Converse. 1986. Changes in soil chemical properties in and around injected liquid manure bands. *Agron. Abstr.* 1986, p. 196. Am. Soc. Agron. New Orleans.

Cooper, J.R., R.B. Reneau, Jr., W. Kroontje, and G.D. Jones. 1984. Distribution of nitrogenous compounds in a Rhodic Paleudult following heavy manure application. *J. Environ. Qual.* 13:189–193.

Cooper, P., and I.S. Cornforth. 1978. Volatile fatty acids in stored animal slurry. *J. Sci. Food Agric.* 29:19–27.

Craswell, E.T. 1978. Some factors influencing denitrification and nitrogen immobilization in a clay soil. *Soil Biol. Biochem.* 10:241–245.

Davidson, E.A., L.F. Galloway, and M.K. Strand. 1987. Assessing available carbon: Comparison of techniques across selected forest soils. *Comm. Soil Sci. Plant Anal.* 18:45–64.

deCatanzaro, J.B., and E.G. Beauchamp. 1985. The effect of some carbon substrates on denitrification rates and carbon utilization in soil. *Biol. Fert. Soils* 1:183–187.

deCatanzaro, J.B., E.G. Beauchamp, and C. Drury. 1987. Denitrification versus dissimilatory nitrate reduction. *Soil Biol. Biochem.* 19:583–587.

Deherain, P.P. 1897. The reduction of nitrate in arable soil. *Compt. Rend. Acad. Sci.* (Paris). Trans. 1974) In: *Microbial Metabolism*, H.W. Doelle, ed., Dowden, Hutchinson & Ross, Stroudsburg, Pa., pp. 233–234.

Delwiche, C.C., and B.A. Bryan. 1976. Denitrification. *Ann. Rev. Microbiol.* 30:241–262.

Egginton, G.W., and K.A. Smith. 1986a. Losses of nitrogen by denitrification from a grassland soil fertilized with cattle slurry and calcium nitrate. *J. Soil Sci.* 37:69–80.

Egginton, G.W., and K.A. Smith. 1986b. Nitrous oxide emission from a grassland soil fertilized with slurry and calcium nitrate. *J. Soil Sci.* 37:59–67.

Elliot, L.F., and T.M. McCalla. 1972. The composition of the soil atmosphere beneath a beef cattle feedlot and a cropped field. *Soil Sci. Soc. Amer. Proc.* 36:68–70.

Eppley, R.W., and J.L. Coatsworth. 1968. Uptake of nitrate and nitrite by *Ditylum brightwellii*–kinetics and mechanisms. *J. Phycol.* 4:151–156.

Evans, D.G., E. Beauchamp, and J.T. Trevors. 1985. Sulfide alleviation of the acetylene inhibition of nitrous oxide reduction in soil. *Appl. Environ. Microbiol.* 49:217–220.

Evans, W.C. 1977. Biochemistry of the bacterial catabolism of aromatic compounds in anaerobic environments. *Nature* 270:17–22.

Ferguson, M., and E.B. Fred. 1909. Denitrification: The effect of fresh and well-rotted manure on plant growth. *Va. Agr. Exp. Sta. Ann. Rpt.* 1908:134–150.

Focht, D.D., and A.C. Chang. 1975. Nitrification and denitrification processes related to wastewater treatment. *Advan. Appl. Microbiol.* 19:153–186.

Gale, P.M. and J.T. Gilmour. 1986. Carbon and nitrogen mineralization from alfalfa (*Medicago sativam* L.): Comparison of aerobic and anaerobic decomposition. *Am. Soc. Agron. Abstr.* p. 179.
Gamble, T.N., M.R. Betlach, and J.M. Tiedje. 1977. Nutritionally dominant denitrifying bacteria from world soils. *Appl. Environ. Microbiol.* 33:926-939.
Garcia, J.-L. 1973. Séquence des produits formes au cours de la dénitrification dans les sols de rizières du Sénégal. *Ann. Microbiol. Inst. Pasteur* 124B:351-362.
Garica, J.-L., F. Pichinoty, M. Mandel, and B. Greenway. 1977. A new denitrifying saprophyte related to *Pseudomonas pickettii*. *Ann. Microbiol.* (Paris) 128A:229-237.
Germon, J.C. 1980. Étude quantitatif de la dénitrification biologique dans le sol a l'aide de l'acétylène. I. Application a différents sols. *Ann. Microbiol.* 131B:69-80.
Germon, J.C., M. Pocket, R. Chaussod and G. Catroux. 1981. Relations entre l'activité dénitrifiante potentielle et les différentes formes de carboné du sol. In: *Colloque humus-azote*, P. Dutil and F. Jacquin, eds., Int. Soil Sci. Soc. and Inst. Nat. Recher. Agron., Reims, France. pp. 338-341.
Gotoh, S., and Y. Onikura. 1971. Organic acids in a flooded soil receiving added rice straw and their effect on the growth of rice. *Soil Sci. Plant Nutr.* 17:1-8.
Gottschalk, G. 1986. *Bacterial metabolism*. Springer-Verlag, NY. 124 pp.
Greenwood, D.J. 1961. The effect of oxygen concentration on the decomposition of organic materials in soil. *Plant Soil.* 14:360-377.
Greenwood, D.J., and H. Lees. 1960. Studies on the decomposition of amino acids in soils. II. The anaerobic metabolism. *Plant Soil.* 12:69-80.
Gross, P.J., J.M. Bremner, and A.M. Blackmer. 1982. A source of error in measurement of denitrification by the acetylene blockage method. *Am. Soc. Agron. Meet. Abstr.* Anaheim, CA, Dec. 3. p. 188.
Guenzi, W.D., and W.E. Beard. 1981. Volatile fatty acids in a redox-controlled cattle manure slurry. *J. Environ. Qual.* 10:479-482.
Guenzi, W.D., W.E. Beard, F.S. Watanabe, S.R. Olsen, and L.K. Porter. 1978. Nitrification and denitrification in cattle manure-amended soil. *J. Environ. Qual.* 7:196-202.
Haider, K., A. Mosier, and O. Heinemeyer. 1985. Phytotron experiments to evaluate the effect of growing plants on denitrification. *Soil Sci. Soc. Am. J.* 49:636-641.
Haider, K., A. Mosier, and O. Heinemeyer. 1987. The effect of growing plants on denitrification at high soil nitrate concentrations. *Soil Sci. Soc. Am. J.* 51:97-102.
Hansen, R.W. and R.E. Kallio. 1957. Inability of nitrate to serve as a terminal oxidant for hydrocarbons. *Science* 125:1190-1199.
Harper, S.H., and J.M. Lynch. 1981. The kinetics of straw decomposition in relation to its potential to produce the phytotoxin acetic acid. *J. Soil. Sci.* 32:627-637.
Hernandez, D., and J.J. Rowe. 1987. Oxygen regulation of nitrate uptake in denitrifying *Pseudomonas aeruginosa*. *Appl. Environ. Microbiol.* 53:745-750.
Hutchinson, H.B. 1918. The influence of plant residues on nitrogen fixation and on losses of nitrate in the soil. *J. Agric. Sci.* 9:92-111.
Ingraham, J.L. 1981. Microbiology and genetics of denitrifiers. In: *Denitrification, nitrification and atmospheric nitrous oxide*, C.C. Delwiche, ed., Wiley, NY. pp. 45-65.
Ishaque, M., and M.I.H. Aleem. 1973. Intermediates of denitrification in the chemoautotroph *Thiobacillus denitrificans*. *Arch. Mikrobiol.* 94:269-282.
Ishaque, M., A. Donawa, and M.I.H. Aleem. 1973. Energy-coupling mechanisms under aerobic and anaerobic conditions in autotrophically grown *Pseudomonas saccharophilia*. *Arch. Biochem. Biophys.* 159:570-579.

Jacobson, S.N., and M. Alexander. 1980. Nitrate loss from soil in relation to temperature, carbon source and denitrifier populations. *Soil Biol Biochem.* 12:501–505.

Jansson, S.L., and F.E. Clark. 1952. Losses of nitrogen during decomposition of plant material in the presence of inorganic nitrogen. *Soil Sci. Soc. Am. Proc.* 16:330–334.

Jenkinson, D.S., and D.S. Powlson. 1976. The effect of biocidal treatments on metabolism in soil. I. Fumigation with chloroform. *Soil Biol. Biochem.* 8:167–177.

Johansson, C., and I.E. Galbally. 1984. Production of nitric oxide in loam under aerobic and anaerobic conditions. *Appl. Environ. Microbiol.* 47:1284–1289.

Knowles, R. 1982. Denitrification. *Microbiol. Rev.* 46:43–70.

Kristjansson, J.K., B. Walter, and T.C. Hollocher. 1978. Respiration dependent proton translocation and the transport of nitrate and nitrite in *Paracoccus denitrificans* and other denitrifying bacteria. *Biochem.* 17:5014–5019.

Laanbroek, H.J., and N. Pfennig. 1981. Oxidation of short-chain fatty acids by sulfate-reducing bacteria in freshwater and in marine sediments. *Arch. Microbiol.* 128:330–335.

Ljungdahl, L.G., and K.E. Eriksson. 1985. Ecology of microbial cellulose degradation. In: *Advances in Microbial Ecology,* vol. 8, K.C. Marshall, ed., Plenum, NY, pp. 237–299.

Lynch, J.M. 1977a. Phytotoxicity of acetic acid produced in the anaerobic decomposition of wheat straw. *J. Appl. Bacteriol.* 42:81–87.

Lynch, J.M. 1977b. Production and phytotoxicity of acetic acid in anaerobic soils containing plant residues. *Soil Biol. Biochem.* 10:131–135.

Lynch, J.M., and K.B. Gunn. 1978. Use of a chemostat to study decomposition of wheat straw in soil slurries. *J. Soil Sci.* 29:551–556.

Lynch, J.M., K.C. Hall, H.A. Anderson, and A. Hepburn. 1980. Organic acids from the anaerobic decomposition of *Agropyron repens* rhizomes. *Phytochemistry* 19:1846–1847.

Marshall, P.R. 1979. Miscible displacement investigations of denitrification in soil. M.Sc. thesis, University of Guelph. 101 pp.

McCarty, P.L., L. Beck, and P. St. Amant. 1969. Biological denitrification of wastewaters by addition of organic material. *Purdue Univ. Eng. Ext. Ser.* 135:1271–1285.

McGarity, J.W., C.M. Gilmour, and W.B. Bollen. 1958. Use of an electrolytic respirometer to study denitrification in soil. *Can. J. Microbiol.* 4:303–316.

Meek, B.D., A.J. MacKensie, T.J. Donovan, and W.F. Spencer. 1974. The effect of large applications of manure on movement of nitrate and carbon in an irrigated desert soil. *J. Environ. Qual.* 3:253–258.

Mishustin, E.N., and N.S. Erofeev. 1964. Nature of the toxic compounds accumulating during the decomposition of straw in oil. *Microbiology* 35:126–129.

Misra, C., D.R. Neilsen, and J.W. Biggar. 1974. Nitrogen transformations in soil during leaching. II Steady state nitrification and nitrate reduction. *Soil Sci. Soc. Am. Proc.* 38:294–304.

Molina, J.A.E., C.E. Clapp, M.J. Shaffer, F.W. Chichester, and W.E. Larson. 1983. NCSOIL, A model of nitrogen and carbon transformations in soil: Description, calibration and behavior. *Soil Sci. Soc. Am. J.* 47:85–91.

Moraghan, J.T., and K.A. Ayotade. 1968. The influence of added organic matter on certain processes occurring in anaerobically incubated soils. *Trans. 9th Int. Cong. Soil Sci.* 4:699–707.

Mosey, F.E. 1983. Mathematical modelling of the anaerobic digestion process: Regulatory mechanisms for the formation of short-chain volatile acids from glucose. *Water Sci. Tech.* 15:209–232.

Mosier, A.R. 1986. Interaction of soluble carbon and soil water content on denitrification. *Am. Soc. Agron. Abstr.* 184 pp.

Myrold, D.D., and J.M. Tiedje. 1985. Establishment of denitrification capacity in soil: Effects of carbon, nitrate and moisture. *Soil Biol. Biochem.* 17:819–822.

Nakajima, M., T. Hayamizu, and H. Nishimura. 1984a. Effect of oxygen concentrations on the rates of denitrification and denitrification in the sediments of an eutrophic lake. *Water Res.* 18:335–338.

Nakajima, M., T. Hayamizu, and H. Nishimura. 1984b. Inhibitory effect of oxygen on denitrification and denitrification in sludge from an oxidation ditch. *Water Res.* 18:339–343.

Nalecz, M.J., R.P. Casey, and A. Azzi. 1986. Use of N,N[1]-decyclohexylcarbodiimide to study membrane-bound enzymes. In: *Methods in Enzymology*, vol. 125, S. Fleischer and B. Fleischer, eds., *Academic Press*, NY, pp. 86–108.

Nelson, D.W., and J.M. Bremner. 1969. Factors affecting chemical transformations of nitrite in soils. *Soil Biol. Biochem.* 1:229–239.

Nelson, D.W., and J.M. Bremner. 1970. Gaseous products of nitrite decomposition in soils. *Soil Biol. Biochem.* 2:203–215.

Nishimura, Y., T. Kamihara, and S. Fukui. 1980. Diverse effects of formate on the dissimilatory metabolism of nitrate in *Pseudomonas denitrificans* ATCC 13867: Growth, nitrite accumulation in culture, cellular activities of nitrate and nitrite reductases. *Arch. Microbiol.* 124:191–195.

Nommik, H. 1956. Investigations on denitrification in soil. *Acta Agric. Scand.* 6:195–228.

Olsen, R.J., R.F. Hensler, and O.J. Attoe. 1970. Effect of manure application, aeration, and soil pH on soil nitrogen transformations and on certain soil test values. *Soil Sci. Soc. Am. Proc.* 34:222–225.

Onodera, I. 1929. *Mem. Morioka Agr. Coll.*, no. 13:176 (cited by Yamane and Sato, 1964).

Patni, N.K., and P.Y. Jui. 1985. Volatile fatty acids in stored dairy cattle slurry. *Agric. Wastes* 13:159–178.

Patten, D.K., J.M. Bremner, and A.M. Blackmer. 1980. Effects of drying and air-dry storage of soils on their capacity for denitrification of nitrate. *Soil Sci. Soc. Am. J.* 44:67–70.

Patrick, W.H. 1960. Nitrate reduction rates in a submerged soil is affected by redox potential. *Trans. 7th Int. Cong. Soil Sci.* (Madison) 2:494–500.

Payne, W.J. 1981. *Denitrification*, Wiley, NY, 214 pp.

Perry, J.J. 1979. Microbial cooxidations involving hydrocarbons. *Microbiol. Rev.* 43: 59–72.

Pfitzner, J., and H.G. Schlegel. 1973. Denitrifikation bei *Hydrogenomonas eutropha* Stam. H16. *Arch. Mikrobiol.* 90:199–211.

Pilot, L., and W.H. Patrick. 1972. Nitrate reduction in soils. Effect of soil moisture tension. *Soil Sci.* 114:312–316.

Rainville, N., and A. Morin. 1985. Changes in the volatile fatty acids content of laboratory stored sterilized and non-sterilized swine wastes. *Microbios.* 42:175–182.

Reddy, K.R., R. Khaleel, and M.R. Overcash. 1980. Carbon transformations in land areas receiving organic wastes in relation to nonpoint source pollution: A conceptual model. *J. Environ. Qual.* 9:434–442.

Reddy, K.R., P.S.C. Ras, and R.E. Jessup. 1982. The effect of carbon mineralization on denitrification kinetics in mineral and organic soils. *Soil Sci. Soc. Am. J.* 46:62–68.

Rhee, G-Y, and G.W. Fuhs. 1978. Wastewater denitrification with one-carbon compounds as energy source. *J. Water Pollut. Control Fed.* 50:2111–2119.

Rolston, D.E., P.S.C. Rao, J.M. Davidson, and R.E. Jessup. 1984. Simulation of denitrification losses of nitrate fertilizer applied to uncropped, cropped, and manure-amended field plots. *Soil Sci.* 137:270–279.

Ryden, J.C., and L.J. Lund. 1980. Nature and extent of directly measured denitrification losses from some irrigated vegetable crop production units. *Soil Sci. Soc. Am. J.* 44: 505–511.

Schennen, U., K. Braun, and H.J. Knackmuss. 1985. Anaerobic degradation of 2-fluorobenzoate by benzoate-degrading, denitrifying bacteria. *Appl. Environ. Microbiol.* 161:321–325.

Schink, B., and N. Pfennig. 1982. Fermentation of trihydroxybenzenes by *Pelobacter acidigallici* gen. nov. sp. nov., a new strictly anaerobic nonspore forming bacterium. *Arch. Microbiol.* 133:195–201.

Schloemer, R., and R.H. Garrett. 1974. Nitrate transport system in *Neurospora crassa*. *J. Bacteriol.* 118:259–269.

Schuman, G.E., and T.M. McCalla. 1976. Effect of short-chain fatty acids extracted from beef cattle manure on germination and seedling development. *Appl. Environ. Microbiol.* 31:655–660.

Sherwood, M.T. 1980. The effects of landspreading of animal manures on water quality. In: *Effluents from Livestock*, J.K.R. Gasser, ed., Applied Science Publishers, Essex, England, pp. 379–392.

Sleat, R., and J.P. Robinson. 1984. The bacteriology of anaerobic degradation of aromatic compounds. *J. Appl. Bacteriol.* 57:381–394.

Smid, A.E., and E.G. Beauchamp. 1976. Effects of temperature and organic matter on denitrification in soil. *Can. J. Soil Sci.* 56:385–391.

Smith, C.J., and W.H. Patrick, Jr. 1983. Nitrous oxide emission as affected alternate anaerobic and aerobic conditions from soil suspensions enriched with ammonium sulfate. *Soil Biol. Biochem.* 15:693–697.

Smith, M.S., and J.M. Tiedje. 1979a. The effect of roots on soil denitrification. *Soil Sci. Soc. Am. J.* 43:951–955.

Smith, M.S., and J.M. Tiedje. 1979b. Phases of denitrification following oxygen depletion in soil. *Soil Biol. Biochem.* 11:261–267.

Sorensen, J., D. Christensen, and B.B. Jorgenson. 1981. Volatile fatty acids and hydrogen as substrates for sulfate-reducing bacteria in anaerobic marine sediment. *Appl. Environ. Microbiol.* 42:5–11.

Stanford, G., R.A. Vanderpol, and S. Dzionia. 1975. Denitrification rates in relation to total and extractable soil carbon. *Soil Sci. Soc. Am. Proc.* 39:284–289.

Stefanson, R.C. 1972. Soil denitrification in sealed soil-plant systems. I. Effect of plants, soil water content and soil organic matter content. *Plant Soil* 7:113–127.

Stevens, R.J., and I.S. Cornforth. 1974. The effect of pig slurry applied to a soil surface on the composition of the soil atmosphere. *J. Sci. Fed. Agric.* 25:1263–1272.

Stouthamer, A.H. (1976). *Yield studies in microorganisms*. Meadowfield Press, Durham, England, 88 pp.

Taylor, B.F. 1983. Aerobic and anaerobic catabolism of vanillic acid and some other methoxyaromatic compounds by a *Pseudomonas* sp. strain PN-1. *Appl. Environ. Microbiol.* 46:1286–1292.

Taylor, B.F., and M.J. Hebb. 1972. The anaerobic degradation of aromatic compounds by a denitrifying bacterium. *Arch. Microbiol.* 83:165–171.

Taylor, B.F., W.L. Campbell, and I. Chinoy. 1970. Anaerobic degradation of the benzene nucleus by a facultatively anaerobic microorganism. *J. Bacteriol.* 102:430–437.

Taylor, B.F., W.L. Hearn, and S. Pincus. 1979. Metabolism of monofluorobenzoates and monochlorabenzoates by a denitrifying bacterium. *Arch. Microbiol.* 122:301-306.

Tenney, F.G., and S.A. Waksman. 1930. Composition of natural organic materials and their decomposition in the soil: V. Decomposition of various chemical constituents in plant materials, under anaerobic conditions. *Soil Sci.* 29:143–160.

Terry, R.E., and J.M. Duxbury. 1985. Acetylene decomposition in soils. *Soil Sci. Soc. Am. J.* 49:90–94.

Tiedje, J.M. 1982. Denitrification. In: *Methods of Soil Analysis, part* 2, A.L. Page, ed., Am. Soc. Agron., Madison, WI, pp. 1011-1026.

Tiedje, J.M., A.J. Sexstone, D.R. Myrold, and J.A. Robinson. 1982. Denitrification: Ecological niches, competition and survival. *Antonie van Leeuwenhoek* 48:569–583.

Timmer-Ten Hoor, A. 1975. A new type of thiosulphate oxidizing, nitrate reducing microorganism: *Thiomicrospira denitrificans* sp. nov. *Neth. J. Sea Res.* 9:343–353.

Timmer-Ten Hoor, A. 1976. Energetic aspects of the metabolism of reduced sulphur compounds in *Thiobacillus denitrificans*. *Antonie van Leeuwenhoek* 42:483–492.

Timmer-Ten Hoor, A. 1977. Denitrificerende Kleurloze Zwavelbacterien. Ph.D. diss., Univ. of Groningen.

Trevors, J.T. 1985. The influence of oxygen concentrations on denitrification in soil. *Appl. Microbiol. Biotechnol.* 23:152–155.

Trevors, J.T., and E. Beauchamp. 1985. Reduction of nitrous oxide (N_2O) by bacterial isolate in the presence of sulfide and acetylene. *J. Microbiol. Meth.* 4:127–131.

Trevors, J.T., and M. E. Starodub. 1988. The influence of oxygen concentrations on denitrification in freshwater sediment. *J. Basic Microbiol.*, 27:387–391.

Verhoeven, W., A.L. Koster, and M.C.A. Van Nievelt. 1954. Studies on true dissimilatory nitrate reduction. III. *Micrococcus denitrificans* beijerinck, a bacterium capable of using molecular hydrogen in denitrification. *Antonie van Leeuwenhoek*. 20:273–284.

Wallace, J.M., and L.F. Elliot. 1979. Phytotoxins from anaerobically decomposing wheat straw. *Soil Biol. Biochem.* 11:325–330.

Wallingford, G.W., L.S. Murphy, W.L. Powers, and H.L. Manges. 1975. Denitrification in soil treated with beef-feedlot manure. *Commun. Soil Sci. Plant Anal.* 6:147–161.

Watanabe, I., and M.R. deGuzman. 1980. Effect of nitrate on acetylene disappearance from anaerobic soil. *Soil Biol. Biochem.* 12:193–194.

Wegener, W.S., H.C. Reeves, R. Rabin, and S.J. Ajl. 1968. Alternate pathways of metabolism of short-chain fatty acids. *Bacteriol. Rev.* 32:1-26.

Widdel, F., and N. Pfennig. 1982. Studies on dissimilatory sulfate-reducing bacteria that decompose fatty acids. II. Incomplete oxidation of propionate by *Desulfobulbus propionicus* gen. nov., sp. nov. *Arch. Microbiol.* 131:360–365.

Wiljer, J., and C.C. Delwiche. 1954. Investigations on the denitrifying process in soil. *Plant Soil* 5:155–169.

Williams, A.G. 1983. Organic acids, biochemical oxygen demand and chemical oxygen demand in the soluble fraction of piggery slurry. *J. Sci. Fed. Agric.* 34:212–220.

Williams, R.J., and W.C. Evans. 1975. The metabolism of benzoate by *Moraxella* species through anaerobic nitrate respiration. *Biochem. J.* 148:1–10.

Woldendorp, J.W. 1963. The influence of living plants on denitrification. *Meded. Ladb. Hoogesch. Wageningen* 63:1–100.

Yamane, I., and K. Sato. 1964. Decomposition of glucose and gas formation in flooded soil. *Soil Sci. Plant Nutr.* 10:38–41.

Yamane, I., and K. Sato. 1968. Initial drop of oxidation-reduction potential in submerged air-dried soils. *Soil Sci. Plant Nutr.* 14:68–72.

Yashura, A., and K. Fuwa. 1977. Odor and volatile compounds in liquid swine manure. I. Carboxylic acids and phenols. *Bull. Chem. Soc. Japan* 50:731–733.

Yeomans, J.C., and E.G. Beauchamp. 1982. Acetylene as a possible substrate in the denitrification process. *Can. J. Soil Sci.* 62:139–144.

Yeomans, J.C., and J. Bremner. 1986. Effects of organic solvents on denitrification in soil. *Am. Soc. Agron. Abstr.* p. 192.

Yoshinari, T., and R. Knowles. 1976. Acetylene inhibition of nitrous oxide reduction by denitrifying bacteria. *Biochem. Biophys. Res. Comm.* 69:705–710.

Young, L.Y. 1984. Anaerobic degradation of aromatic compounds, In: *Microbial degradation of organic compounds*, D.T. Gibson, ed., Marcel Dekker, NY, pp. 487–523.

Young, L.Y., and A.C. Frazer. 1987. The fate of lignin and lignin-derived compounds in anaerobic environments. *Geomicrobiol. J.* 5:261–293.

Integrated Nitrogen Management in Irrigated Rice*

S.K. De Datta† and R.J. Buresh‡

I. Introduction

More than 90% of the rice (*Oryza sativa* L.) in the world is produced in Asia (IRRI, 1986b). Within Asia, rice production has been increasing an average 2.7% annually—slightly faster than population growth. The increased production is

*Joint contribution from the International Rice Research Institute (IRRI), P.O. Box 933, Manila, Philippines, and the International Fertilizer Development Center (IFDC), P.O. Box 2040, Muscle Shoals, AL 35662, USA.
†Principal Scientist and Head, Department of Agronomy, IRRI.
‡Soil Scientist, IFDC and Visiting Scientist, IRRI.

Advances in Soil Science, Volume 10

primarily due to higher yields per hectare rather than increased land area under cultivation. The adoption of modern varieties, increased irrigation facilities, and more fertilizer use have been credited for the yield increases. An estimated 24% of the increase in Asian rice production from 1965 to 1980 was attributed to use of fertilizer, mainly N (Barker *et al.*, 1985).

Asia consumed 39% of the world's N fertilizer in 1985/1986 (FAO, 1987) as compared to only 24% in 1973 (FAO, 1974). The annual increase in consumption rates in China, India, Pakistan, and Indonesia, among the Asian countries, was more than 14% for most of the period from 1973 to 1983 (Stangel and De Datta, 1985). Nitrogen represents approximately 70% of the total fertilizer nutrient (N + P_2O_5 + K_2O) consumption in Asia (FAO, 1987), and approximately 60% of the N fertilizer consumed in Asia is used on rice (Stangel and De Datta, 1985).

Urea accounts for more than 75% of the total N fertilizer consumption in Asia. It is the major N fertilizer for rice, although ammonium bicarbonate is also extensively used in China. In 1985, China produced 11.4 million tons of N, 45% as urea and 55% as ammonium bicarbonate (Maene *et al.*, 1987). China is the second largest N producer following the USSR (14.0 million tons in 1985/1986). The United States (9.4 million tons in 1985/1986) and India (4.3 million tons in 1985/1986) are the third and fourth largest N producers, respectively (FAO, 1987). Urea production in 1985/1986 in India reached 3.4 million tons of N from 26 fertilizer plants (Maene *et al.*, 1987).

Despite this impressive growth in urea production and use, research over the past 20 to 30 years has shown that N fertilizers are generally inefficient, with less than 40% of the applied N normally used by lowland rice (De Datta *et al.*, 1968; Craswell and Vlek, 1979). In recent years, a major research effort to devise efficient management practices and modified forms of urea has been advocated (McCune, 1982). Basic N transformation processes and alternative management practices to improve N fertilizer use efficiency in flooded rice were reviewed in a recent publication (De Datta and Patrick, 1986).

During the past 20 years, the use of green manure crops, animal manures, and composts, which traditionally were important sources of nutrients, declined substantially as chemical fertilizers became increasingly available. In some countries, chemical fertilizer prices were subsidized, thereby enabling farmers to apply production-maximizing doses.

Organic materials remain an important source of nutrients in China, although the proportion of chemical fertilizer is steadily increasing. In 1977, China consumed 8.5 million tons of nutrient from chemical fertilizers and 23 million tons of nutrient from organic sources (Tang and Stone, 1980). By 1985/1986, the consumption of nutrient from chemical fertilizers had increased to 16.9 million tons (FAO, 1987). At the end of the 1970s, the area in China planted to green manure crops, mainly milk vetch (*Astragalus sinicus* L.), was 10 million ha; 90% of it in rice fields. By the mid-1980s the area decreased to 6 million ha (Chen, 1988). In Japan, the use of green manure crops, which reached a maximum of 505,000 ha in 1934, declined to 4,500 ha by 1980 (Ishikawa, 1988).

An issue of increasing concern is the sustainability of soil productivity as lands are called upon to produce higher yields from a single crop and higher total yields under intensive cropping systems. Hence, sound strategies for increasing fertilizer use efficiency and for integrated N management in rice need to be developed. The greatest payoffs from increased fertilizer use efficiency and integrated N management in rice will occur in the irrigated areas of Asia because of the dominance of Asia in global rice production and the production of over 70% of the world's rice under irrigated conditions. This chapter summarizes recent research findings on irrigated-rice-production systems in Asia.

II. Efficiency of Nitrogen Fertilizer

Organic sources of N in rice-production systems include animal wastes, crop residues, green manures, and urban wastes (Heese, 1984). They can substitute for chemical fertilizers, but to sustain high yields of irrigated rice some chemical N fertilizer must be used with the organic N fertilizers. Therefore, increased crop utilization and reduced losses of chemical N fertilizer will remain an important goal for effective N management. The reduction of N losses may gain importance as concerns over groundwater and environmental pollution increase.

Poor utilization of chemical N fertilizer by rice is thought to be largely due to N losses from the soil/plant system. Ammonia volatilization, denitrification, runoff, and leaching are major mechanisms of loss. Obviously, the nature of the N fertilizer, the application method, and field-level conditions all affect fertilizer efficiency. The percentage of added N recovered by rice can be increased by deep placement of chemical N fertilizer in soil or topdressing chemical N fertilizer at later growth stages (Craswell and Vlek, 1979; Prasad and De Datta, 1979).

The primary goal of improved N management practices should be to maximize N uptake at critical growth stages and to minimize transformation processes that lead to losses or temporary losses (e.g., ammonium fixation and immobilization) of N from soil/water systems (De Datta, 1987). It is critical to ensure that N absorbed by the plant is used for grain production, which has been designated as agronomic efficiency. Agronomic efficiency refers to the amount of rice produced per kilogram of nutrient applied, and under different environmental conditions it is influenced by the interaction of climatic, soil, and plant factors (Craswell and Godwin, 1984).

III. Ammonia Volatilization Loss

Ammonia volatilization is a major loss mechanism that affects the efficiency of urea and other N fertilizers in irrigated lowland rice. The magnitude of ammonia loss depends on wind speed, temperature, rainfall, ammoniacal-N (NH_4^+ + NH_3) concentration, pH, and cation exchange capacity (Freney *et al.*, 1983). In Asian rice fields, the high floodwater ammoniacal-N concentrations following N application, high temperatures common in the tropics, and elevated floodwater pH resulting from photosynthetic activity create a favorable environment for ammonia loss (Fillery *et al.*, 1986b; Vlek and Craswell, 1981). Vlek and

Craswell (1981) explained that aqueous ammonia content in floodwater increases about tenfold per unit increase in pH over the pH range 7.5 to 9.0. At constant total ammoniacal-N concentration and pH, aqueous ammonia increases almost linearly with increasing temperature, which results in nearly a fourfold increase in the range from 10 to 40°C. In field systems, floodwater pH displays a diurnal pattern seemingly synchronized with the cycles of photosynthesis and net respiration or the depletion and addition of CO_2 to floodwater (Mikkelsen *et al.*, 1978).

A. Measurement of Ammonia Volatilization

Most measurements of ammonia loss from flooded rice soils have been made with static or forced-air-exchange enclosures (Savant and De Datta, 1982; Fillery *et al.*, 1984). Nonetheless, there is considerable uncertainty regarding the interpretation of results from such artificial enclosures because they disrupt the environmental and soil processes influencing gas exchange (Denmead, 1983).

Micrometeorological techniques have recently enabled scientists to take direct field measurement of ammonia loss without disrupting the natural environmental conditions (Denmead, 1983). The integrated horizontal flux technique requires a circular plot of approximately 25-m radius and measurement of time-averaged wind speed and atmospheric ammonia concentration at five or more levels above the lowland rice field. A background profile measurement of time-averaged atmospheric ammonia concentration is necessary to calculate net ammonia flux from the circular area. The technique is labor intensive, and it requires an electrical power supply together with considerable instrumentation in the field.

Wilson *et al.* (1982) proposed a simplification of the integrated horizontal flux technique in which wind speed and atmospheric ammonia concentration are measured at a single height. This simplified horizontal transport technique has been successfully verified against the integrated horizontal flux technique in a grass pasture (Wilson *et al.*, 1983) and irrigated lowland rice fields (Freney *et al.*, 1985; Fillery and De Datta, 1986). This simplified technique requires fewer anemometers, flow meters, and ammonia traps; but electrical power and a circular field plot of approximately 25-m radius are still needed.

Scientists at the Commonwealth Scientific and Industrial Research Organization (CSIRO) in Australia have continued to develop additional simple methods for nondisruptive field measurement of ammonia loss. Freney *et al.* (1985) presented a bulk aerodynamic technique that is based on the measurement of floodwater parameters (ammoniacal-N, pH, and temperature) and wind speed at a single height. Unlike the simplified horizontal transport technique, the bulk aerodynamic technique can be used on small field plots; hence it permits the comparison of ammonia loss from several fertilizer treatments. This technique is less accurate than the simplified horizontal transport technique because transport processes in the water and atmosphere controlling ammonia emission can vary diurnally and with floodwater environment (Leuning *et al.*, 1984; Freney *et al.*, 1985). The technique is only applicable in environments with standing water.

Leuning *et al.* (1985) described an integrative ammonia sampler that enables measurement of ammonia fluxes without the necessity of measuring wind speed

Table 1. Summary of recent measurements of ammonia volatilization by the micrometeorological technique and total N loss by the ^{15}N balance technique for urea applied to transplanted lowland rice

Urea application method	Reference	Site[a]	Urea-N applied	Ammonia volatilization	Total N loss	Estimated denitrification
			kg/ha	% of applied N		
Basal incorporation	Fillery *et al.*, 1986c	Muñoz	80	15	18	3
	Fillery *et al.*, 1986c	Los Baños	60	13	26	13
	Cai *et al.*, 1986	Dan Yang	90	9	48	39
Broadcast at 10–21 DT[b]	Fillery *et al.*, 1986c	Muñoz	80	47	45	0
	Fillery and De Datta, 1986	Muñoz	58	36	41	5
	Fillery *et al.*, 1986c	Los Baños	60	27	60	33
	Fillery and Byrnes, 1984	Los Baños	60	53	55	2
	IRRI, 1986a	Mabitac	80	31[c]	55	24
	IRRI, 1987	Calauan	80	54	64	10
	IRRI, 1988	Calauan	80	9	43	34
	IRRI, 1987	Aguilar	80	30	49	19
	Bidin *et al.*, 1987[d]	Bumbong Lima	40	7	39	32
Broadcast at 5–7 DBPI[b]	Fillery *et al.*, 1986c	Muñoz	40	11	13	0

[a] All sites are in the Philippines except for Dan Yang and Bumbong Lima, which are in the People's Republic of China and Malaysia, respectively.
[b] DT designates d after transplanting; DBPI designates d before panicle initiation.
[c] Estimated by the simplified horizontal transport technique.
[d] The methods of rice crop establishment and timing of urea application were not indicated.

and atmospheric ammonia concentration. The sampler is placed at one height in the center of circular plots. It requires no electrical power supply and is simple to use. Unlike the bulk aerodynamic technique, it can be used for measurement of ammonia loss from treatments without continual standing floodwater.

B. Magnitude of Ammonia Loss

The first direct measurements of ammonia loss with the micrometeorological technique in tropical irrigated rice fields were made by Freney *et al.* (1981). Their study was conducted with ammonium sulfate applied to a puddled lowland rice field in the Philippines. Ammonia loss accounted for 5% of the ammonium sulfate-N broadcast and incorporated (80 kg N/ha) before transplanting and 11% of the ammonium sulfate-N broadcast (40 kg N/ha) into floodwater at panicle initiation. Subsequent field measurements of ammonia loss have focused primarily on urea.

Table 1 summarizes recent micrometeorological measurements of ammonia loss from urea in Asian irrigated rice environments. Ammonia loss of applied urea-N ranged from 7 to 54%. In the Philippines, where most of the measurements were conducted, ammonia loss tended to be greater from urea broadcast into floodwater at 10 to 21 d after transplanting (DT) than from urea either broadcast and incorporated before transplanting or broadcast at 5 to 7 d before panicle initiation (DBPI). The high measured ammonia loss from urea broadcast at 10 to 21 DT (Table 1), a common farmers' practice in the Philippines and many areas of Asia, highlights the importance of ammonia volatilization as a loss mechanism. The high ammonia loss has been attributed to high concentrations of ammoniacal-N following urea application, high floodwater pH, high temperature, and high wind speed (Fillery and Vlek, 1986).

Total N loss in Table 1 was obtained with the ^{15}N balance technique in which metal microplot borders are used to prevent runoff loss. In all studies at Muñoz, nearly all the N loss was as ammonia volatilization. Fillery *et al.* (1986c) attributed this to high wind speed, which favored rapid air exchange and loss of ammonia. Fillery *et al.* (1986c) found lower ammonia loss, but not total loss, from broadcast urea at Los Baños (27%) than Muñoz (47%). They concluded that the lower ammonia loss at Los Baños was due to lower wind speed, and they speculated that low wind speed and high soil nitrification potential at Los Baños, unlike Muñoz, created conditions favorable for N loss by nitrification/denitrification. However, in a study 2 years later on the same soil and in the same season at Los Baños, Fillery and Byrnes (1984) found double the ammonia loss (53%) and virtually no N loss by other mechanisms. These results suggest that the magnitude of ammonia loss and its importance relative to other loss mechanisms can vary greatly from year to year.

Ammonia loss also varied greatly between 2 years at Calauan (54% and 9%) (Table 1). The cause for this variation is not completely clear, although the 9% measured ammonia loss at this site might be erroneously low because of seepage of floodwater containing fertilizer-N from the circular plot immediately following urea application.

Ammonia loss from urea broadcast and incorporated before transplanting was less at Dan Yang, Jiangsu Province, China (9%) than at two sites in the Philippines (13% and 15%) (Table 1). Moreover, total N loss and hence loss by another mechanism, presumably nitrification/denitrification, was much greater at Dan Yang. Cai *et al.* (1986) attributed the low ammonia loss at Dan Yang to low-incident solar radiation and low floodwater pH at the experimental site.

Bidin *et al.* (1987) similarly reported low ammonia loss (7%) from urea in Malaysia, even though total loss of applied urea-N was substantial (39%). Bidin *et al.* (1987) concluded that under the production systems and environmental

Table 2. Effects of timing, application method, and floodwater depth on loss of 53 kg urea-N/ha applied to lowland rice at Mabitac, Philippines

Urea application			N loss		
Timing	Method	Floodwater depth	Ammonia volatilization	Total	Denitrification[a]
		m	% of applied N		
Basal	Incorporation	0	11	33	22
Basal	Incorporation	0.05	42	54	12
10 DT[b]	Broadcast	0.05	48	60	12

[a] Denitrification was estimated by difference between unaccounted-for ^{15}N in a ^{15}N balance and ammonia loss obtained by the simplified bulk aerodynamic technique.
[b] DT designates d after transplanting.
Source: Unpublished IRRI/CSIRO collaborative research.

regimes in peninsular Malaysia, ammonia loss may not be the major loss mechanism in lowland rice. The limited available micrometeorological measurements of ammonia loss indicate that ammonia volatilization loss varies greatly from site to site depending on the prevailing soil, water, and atmospheric parameters following N application.

The simplified bulk aerodynamic technique has been used in collaborative IRRI/CSIRO research to simultaneously estimate ammonia loss for different fertilizer management practices. The technique has been useful in demonstrating the strong effects of urea timing and water management on ammonia loss (Table 2). Incorporation of urea before transplanting with no standing floodwater rather than 0.05-m floodwater reduced urea-N loss as ammonia from 42 to 11% and reduced total N loss from 54 to 33%. The reduced ammonia loss when urea is incorporated without floodwater was attributed to more thorough incorporation of urea into the puddled soil and less escape of urea-N and subsequently ammoniacal-N to the soil surface and floodwater where it is prone to ammonia loss.

By using a combination of the bulk aerodynamic technique with the ^{15}N balance technique, researchers can estimate denitrification by difference. However, a limitation of this approach is that ammonia loss can only be estimated when the soil is flooded. In the basal incorporation treatment with no floodwater reported in Table 2, ammonia loss measurements could not begin until the soil was flooded 2 d after N application. Therefore, ammonia loss prior to flooding was not measured and would be erroneously attributed to denitrification loss.

Wetselaar *et al.* (1986) measured 59 to 65% total N loss by the ^{15}N balance technique when urea was broadcast at 90 kg N/ha into floodwater at Sukamandi, West Java, Indonesia. Approximations of ammonia loss with the bulk aerodynamic technique using measured pNH_3 and the relationship of wind speed and pNH_3 to ammonia flux published by Fillery *et al.* (1984) suggested that all gaseous N loss was as ammonia.

Cai *et al.* (1986) compared ammonia and total N loss for urea and ammonium bicarbonate, an important N source in China. Ammonia loss from basal broadcast and incorporated fertilizer was greater for ammonium bicarbonate (18%) than for urea (9%). Total N loss was also greater for ammonium bicarbonate (60%) than for urea (48%). Cai *et al.* (1986) concluded that denitrification represented the difference between total N loss and ammonia loss and hence was the major mechanism of N loss for both ammonium bicarbonate (42%) and urea (39%).

C. Floodwater Nitrogen Dynamics

As ammonia loss is directly related to partial pressure of ammonia, pNH_3, at a given wind speed (Fillery *et al.*, 1984), researchers have used pNH_3 as a relative assessment of ammonia loss. Buresh (1987) used pNH_3 measurements to illustrate that diammonium phosphate and ammonium phosphate sulfate (16-20-0) are prone to high ammonia loss within the first 3 d after application to rice fields, whereas ammonia loss from urea is more gradual, extending over 7 d. De Datta *et al.* (1987b) used pNH_3 measurements following basal incorporation of urea with 0.05-m standing water to show the substantial release of ammoniacal-N to the floodwater where it was prone to ammonia loss.

Rao (1987) in a field examination of granular urea and coated urea observed that total ammonia loss, measured with a semi-open chamber containing an enclosed acid trap, correlated (r = 0.92) with peak levels of ammoniacal-N and pNH_3 at 2 d after N application. Plant N at harvest was negatively correlated (r = 0.96) with ammoniacal-N in floodwater at 2 d after fertilization. The potential applications of single time measurements of floodwater N to assess N loss merit further investigation. Nonetheless, this technique would likely not be valid for treatments with different patterns of ammoniacal-N accumulation in floodwater such as urea with and without a urease inhibitor. Buresh *et al.* (1988c), for example, showed that the peak levels of floodwater ammoniacal-N and pNH_3 with urea and urea containing a urease inhibitor differ by several days. Moreover, because sites can differ in wind speed, floodwater ammoniacal-N or pNH_3 without a concomitant measure of wind speed would not provide a valid assessment of ammonia loss among sites.

IV. Denitrification Loss

Whereas recent research has directly determined the magnitude of ammonia loss under field conditions, very little is known about the magnitude of denitrification loss following N application in Asian rice environments. The direct field measurement of denitrification, although highly desirable, has been hampered by lack of suitable methodology (Hauck, 1986).

A. Measurement of Denitrification

The N_2 produced from denitrification is difficult to measure against a background of 78% N_2 in the atmosphere. Consequently, micrometeorological methods to measure N_2 flux are not available. Prior measurements of N_2O flux in flooded rice fields (Freney *et al.*, 1981; Smith *et al.*, 1982) do not provide a quantitative measure of denitrification loss because the N_2O/N_2 ratio of evolved N gases is not known. The acetylene inhibition technique, which has been widely used on upland soils, does not appear to be applicable to flooded soil environments because acetylene might not be rapidly dispersed throughout saturated soil (Keeney, 1986). Moreover, acetylene is reportedly decomposed in anaerobic soil (Watanabe and de Guzman, 1980), and a significant fraction of the N_2O formed in the presence of acetylene will be dissolved in the soil and overlying floodwater (Moraghan and Buresh, 1977).

The most suitable technique for direct field measurement of denitrification in flooded soils appears to be addition of a highly ^{15}N-enriched source to a microplot and subsequent collection of evolved ^{15}N-enriched gases in a confined atmosphere. The evolved ^{15}N-enriched gases normally are not in isotopic equilibrium with atmospheric N_2 (Hauck *et al.*, 1958). Therefore, either the ion signals for mass-to-charge ratio of 28, 29, and 30 in the collected N_2 gas must be measured (Rolston *et al.*, 1978; Siegel *et al.*, 1982; Mosier *et al.*, 1986) or the N isotopes in the collected N_2 gas must be randomized to allow conventional analysis based on the 29/28 ion signal ratio. Craswell *et al.* (1985) reported a high-voltage electric arc technique to randomize the distribution of the N isotopes. Strong *et al.* (1987) showed that this arc technique determined the ^{15}N content of ^{15}N-enriched N_2O as well as N_2 in air samples.

Buresh and Austin (1988) recently used a floating chamber technique to directly collect $(N_2 + N_2O)$-^{15}N evolved from ^{15}N-labeled urea applied to flooded soil. Their method permitted the measurement of denitrification in a more natural environment than did the method of Iimura and Ito (1984) in which the chamber was permanently sealed into lowland rice soil and the chamber atmosphere was replaced with an 80% He and 20% O_2 gas mixture. Table 3 summarizes recent measurements of $(N_2 + N_2O)$-^{15}N flux from urea-^{15}N made with the technique of Buresh and Austin (1988). At all sites and application methods, only 1% or less of the added urea-N was recovered as $(N_2 + N_2O)$-^{15}N during the 17 to 20 d following urea application. Total N loss determined as unaccounted-for ^{15}N by the ^{15}N balance ranged from 26 to 46% of the applied N in these studies (Table 3). Total N loss presumably represents gaseous N loss as microplot borders prevented runoff and leaching was negligible on these puddled soils. The results clearly indicate that $(N_2 + N_2O)$-N flux represented only a very small fraction of the total gaseous N loss.

The 0.1% loss as $(N_2 + N_2O)$-N at Calauan (Table 3) differs dramatically from the 10 and 34% denitrification loss estimated by difference for the same application method at this site (Table 1). Buresh and Austin (1988) recognized that the measurement of $(N_2 + N_2O)$-N flux underestimated denitrification loss because the direct recovery of $(N_2 + N_2O)$-^{15}N from added ^{15}N-labeled NO_3^- was only 41 to 73% of the total ^{15}N loss when runoff and leaching were prevented. Recent laboratory studies by Lindau *et al.* (1988) and Holt *et al.* (1988) indicated that ^{15}N-labeled gases formed by denitrification can be entrapped within flooded soil. Katyal *et al.* (1989) proposed a technique for measuring entrapped ^{15}N-labeled

Table 3. Summary of recent measurements of $(N_2 + N_2O)$-N flux from urea applied to transplanted lowland rice in the Philippines

Urea application method	Reference	Site	Urea-N applied	Recovery of $(N_2 + N_2O)$-N	Total N loss
			kg/ha	— % of applied N —	
Basal incorporation, 0.02-m water	Buresh and Austin, 1988	Muñoz	58	1.1	40
Basal incorporation, no water	Buresh and Austin, 1988	Muñoz	58	0.2	26
Broadcast at 11–15 DT[a]	Buresh and Austin, 1988	Muñoz	44	0.5	46
	De Datta *et al.*, 1987a	Calauan	80	0.1	45
	John, 1987	Los Baños	29	0.5	41

[a] DT designates d after transplanting.

gases, and field studies are in progress to determine whether this technique together with the measurement of $(N_2 + N_2O)$-^{15}N flux will result in quantification of denitrification. Quantification of denitrification might also necessitate placement of chambers over the rice plants because a considerable fraction of soil gases emitted to the atmosphere can be transferred through rice plants (Seiler *et al.*, 1984; Reddy *et al.*, 1987).

B. Magnitude of Denitrification

Recent indirect estimates of denitrification from the difference between unaccounted-for ^{15}N in a ^{15}N balance and measured or estimated ammonia loss are given in Tables 1 and 2. Estimated denitrification loss ranged from 0 to 39% of the applied urea. In general, the magnitude of denitrification and ammonia loss varied more among experiments than did the total N loss. This observation has led researchers (Fillery and Vlek, 1986; Simpson and Freney, 1986) to speculate that ammonia loss and denitrification loss are concurrent, complementary loss mechanisms in flooded soil. They argued that under conditions of high ammonia accumulation and wind speed, ammonia loss predominates, whereas under conditions of low ammonia accumulation and wind speed, nitrification/denitrification loss predominates. Thus, total gaseous N losses tend to remain high and more constant from year to year and from site to site than either ammonia loss or denitrification loss.

The importance of denitrification in puddled, lowland rice soils remains an enigma:

1. The high denitrification losses estimated by difference (Tables 1 and 2) have not been confirmed by high directly measured losses of N_2 and N_2O (Table 3).
2. The denitrification losses estimated by difference necessitate a daily nitrification/denitrification rate as high as 2.6 kg N/ha/d (Fillery *et al.*, 1986c). However, the highest nitrification rate reported by Watanabe *et al.* (1981) after urea application to flooded soil was only 1.2 kg N/ha/d.
3. Floodwater NO_3^- and NO_2^- normally remains very low following urea application in puddled soil. Freney *et al.* (1981) observed no significant buildup of soil or floodwater NO_3^- following urea application. John (1987) reported peak floodwater (NO_3^- + NO_2^-)-N levels of less than 8 mg/m² following broadcast application of urea at 15 DT. This corresponded to less than 0.3% of the applied N and is lower than reported floodwater (NO_3^- + NO_2^-)-N levels following urea application in direct-seeded rice on nonpuddled soil. Peak amounts of floodwater (NO_3^- + NO_2^-)-N in g/m² reported by Smith *et al.* (1982) in Louisiana and Simpson *et al.* (1984) in Australia were approximately five and fifty times greater, respectively.

Integrated use of organic N sources with chemical fertilizers might be speculated to enhance denitrification loss when denitrification is limited by available C. Nonetheless, John (1987) reported that incorporation of cowpea, *Vigna unguiculata* (L.) Walp., green manure (dry wt = 2.5 Mg/ha, C/N = 15) at 15 d before transplanting had no effect on (N_2 + N_2O)-N flux from urea broadcast 30 d later at 15 DT. Flux in the presence and absence of green manure was only 0.5% of the added urea-N, whereas total N loss, presumably as gases, was 40%. John (1987) concluded that denitrification in the puddled soil at this Los Baños, Philippines, site was limited by supply of nitrate rather than by availability of C.

Organic N fertilizers can be prone to gaseous N loss. Biswas (1988) reported 29% loss of 90 kg N/ha incorporated as ^{15}N-enriched *Sesbania rostrata* green manure immediately prior to rice transplanting. The corresponding N loss from 90 kg urea-N/ha applied by the recommended split was 28%. Since pNH_3 was substantially lower following the green manure than the urea application, Biswas (1988) speculated that gaseous N loss other than ammonia volatilization accounted for at least part of the ^{15}N loss from *S. rostrata* green manure.

V. Management Agenda to Increase Nitrogen Use Efficiency

In recent years, a deeper understanding of the mechanisms causing poor N utilization has helped researchers to develop cultural practices that improve N fertilizer use efficiency in lowland rice. Research results suggest that it is wasteful to apply N in the floodwater between transplanting and early tillering, which is common among farmers in Southeast Asia. Early N uptake is essential for high tiller production (De Datta, 1981).

The selection of more N efficient rice genotypes is now recognized as a way to increase N use efficiency (Broadbent *et al.*, 1987; De Datta and Broadbent,

1988). Rice genotypes have been identified that produce more grain per unit of N fertilizer input than "IR36," which is grown on extensive areas in Asia, and "IR50," which is grown in parts of India. Advanced breeding lines can now be routinely screened and selected for their ability to produce the most grain with the least input of N fertilizer (De Datta and Broadbent, 1988).

A. Urea and Modified Urea Materials

Urea is likely to remain the leading N source for rice until the end of the century as attested by the large potential for new urea capacity in Asia where rice is primarily grown. However, some changes in the form of urea can be expected. Most urea is produced and sold as dry prills, but recently production shifted toward that of granular urea. New granular urea production facilities are either built, under construction, or planned in Malaysia, Thailand, and Mexico (Stangel, 1985).

1. Urease Inhibitors

Urease inhibitors offer the attractive possibility of reducing ammonia loss through addition of a chemical to urea during the fertilizer production stage without requiring a change in the farmers' existing fertilizer application practice. In theory, a urease inhibitor would retard urea hydrolysis, thereby preventing ammonia buildup in the floodwater. The addition of only a small quantity of the inhibitor to urea, approximately 0.02 g/g urea or less, would not appreciably increase transportation costs.

Amending urea with phenyl phosphorodiamidate (PPD), which is probably the most widely tested urease inhibitor for rice, has been shown to lower N loss by ammonia volatilization (Fillery and De Datta, 1986), but it has not consistently increased grain yield (Fillery *et al.*, 1986a). In recent collaborative IRRI/IFDC field studies, N-(n-butyl) thiophosphoric triamide (NBPT) was much more effective than PPD in reducing pNH_3 at two sites in the Philippines (Buresh *et al.*, 1988c). However, amendment of urea with NBPT significantly increased grain yield at only one of the two sites (Buresh *et al.*, 1988d).

The full potential of inhibitors to increase the yield response of rice to urea has been uncertain as no available chemical completely eliminated gaseous N loss. Knowledge on the total increase in grain yield possible with inhibitors would facilitate an assessment on the merits of inhibitors relative to other strategies to increase N efficiency, such as urea coatings and urea deep placement. In a recent study, Buresh *et al.* (1988b) defined an ideal inhibitor as a chemical additive to urea that completely eliminated all gaseous loss of applied urea-N while having no effect on rice other than increasing the availability of added N. Thus, they assumed that unaccounted-for ^{15}N in ^{15}N balance studies at multiple urea rates, in the absence of leaching and runoff loss, represented the potential savings in urea-N with an ideal inhibitor. Potential increases in grain yield with complete elimination of gaseous N loss were calculated as the difference between estimated yields, assuming no N loss and estimated yields with existing N loss

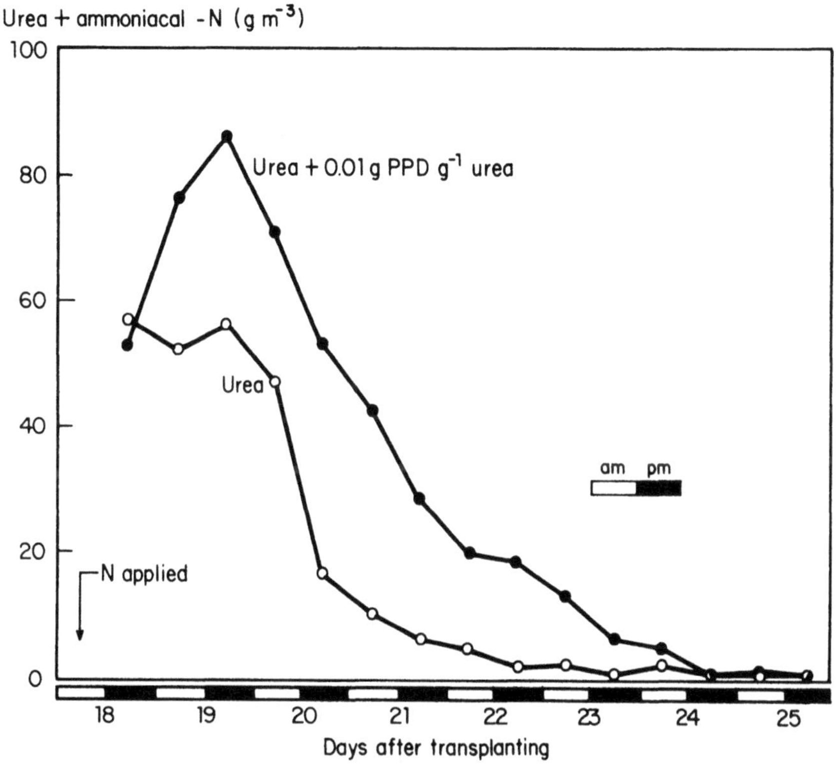

Figure 1. Effect of urease inhibitor, phenyl phosphorodiamidate (PPD), on floodwater (urea + ammoniacal)-N following application of 58 kg urea-N ha^{-1} at 18 d after transplanting. (Adapted from Castillo, 1985).

(Blackmer, 1986). In a 2-year study at one site in the Philippines, in which total grain yield response to applied urea was approximately 3 Mg/ha, Buresh *et al.* (1988b) concluded that an ideal inhibitor could increase yield by a maximum of 0.4 to 0.5 Mg/ha or 6 to 8%.

Although urease inhibitors are effective in reducing the accumulation of ammoniacal-N in floodwater following urea application, they tend to delay the disappearance of applied urea from floodwater (Buresh *et al.*, 1988c; Fillery *et al.*, 1986a). Hence, the concentration of floodwater (urea + ammoniacal)-N following broadcast application of urea to floodwater tends to be increased with the use of a urease inhibitor (Figure 1). Wetselaar *et al.* (1986) made a similar observation in West Java, Indonesia, and concluded that urease inhibitors might act to increase the runoff loss of N in environments where water flows out of the rice field soon after fertilization.

2. Coated Urea

Modification of urea with coatings offers the possibility of reducing N loss and increasing the utilization of added N through slow release of N to rice. Basal

Table 4. Effect of sulfur coatings on the savings in urea at one site in the Philippines

N source	Composition N	Composition S	Composition Sealant	Composition Conditioner	Total coating	Urea granule diameter	7-d release rate in water	Urea savings[a] at 5 Mg/ha	Urea savings[a] at 6 Mg/ha
	%					mm	%	%	
SCU-21	21	15.6	NA[b]	NA	20.1	2.0	21	45	48
SCU-36	39	14.8	0	0	14.8	4.4	36	29	34
SCU-43	40	9.1	1.2	2.1	12.4	4.4	43	48	52
SCU-70	41	8.9	0.6	2.2	11.7	4.4	70	36	39
SCU-100	41	9.8	0	0	9.8	4.4	100	25	29

[a] Urea savings $= \left[\frac{N_u - N_{scu}}{N_u}\right] 100$

where N_u = urea-N required for a desired yield,
N_{scu} = SCU-N required for the desired yield.
[b] NA = not available.
Source: From IRRI, 1987.

incorporation of sulfur-coated urea (SCU) has been shown to be effective for rice (De Datta and Gomez, 1981). However, SCU is expensive to produce from approximately 2-mm-diameter granular urea because a large total coating (sulfur, sealant, and conditioner) of approximately 0.2 g/g SCU is required. One way of decreasing the quantity of coating per weight of urea is to increase the urea granule size (Parish *et al.*, 1980). Recent collaborative IRRI/IFDC field studies demonstrated that the total coating of SCU could be reduced to 0.12 g/g SCU for 4.4-mm-diameter granular urea without affecting grain yield and savings in urea obtained with sulfur coating (Table 4). The savings in urea obtained with SCU fertilizers (Table 4) are greater than those observed in the Philippines for the urease inhibitors PPD and NBPT (Buresh *et al.*, 1988d). Research and development are continuing on identification and testing of suitable alternative coatings that would be only 0.03 g/g fertilizer or less (IFDC, 1988).

3. Urea Deep Placement

Deep placement of urea in lowland rice fields is widely recognized as an effective management practice for transplanted rice, except on soils with a high percolation rate (Katyal *et al.*, 1985; Savant and Stangel, 1985). Basal deep placement of urea is also superior to split applications of prilled urea in broadcast-seeded flooded rice (De Datta *et al.*, 1988). Because hand-point placement of urea as urea supergranules (USG) is labor intensive, recent research has focused on the development and evaluation of less labor-intensive methods to facilitate deep placement. The methods proposed and evaluated include: (1) pneumatic injection of prilled or granular urea (Prins *et al.*, 1984); (2) mechanical deep placement of prilled urea (IRRI, 1986a); (3) mechanical deep placement of liquid urea

(Schnier *et al.*, 1988); (4) injection as a mud/urea slurry, which is being developed by the Vietnam Institute of Agricultural Engineering and Mechanization, Hanoi; and (5) a USG dispenser to facilitate simultaneous line transplanting and deep placement (Savant, 1988). Considerable scope still exists for the development of less labor-intensive, economical methods of deep placement. Moreover, better delineation is needed of environments where deep placement has a distinct, consistent advantage over other management practices.

B. N Timing and Water Management

De Datta *et al.* (1979) examined the relative contribution of three biological constraints (fertilizer, weed control, and insect control) to the gap between potential yield, with recommended management practices, and actual farm-level yield at three provinces in the Philippines. They concluded that over the 2 to 4 years of the study, improper management and insufficient amounts of fertilizer were a major constraint, which accounted for approximately 1.0 and 0.5 Mg/ha reductions in farm-level yield in the dry and wet seasons, respectively. Fertilizer management could be improved by: (1) basal incorporation of the first N dose rather than broadcast application at 10 to 21 DT when N losses are extremely high (Table 1) and (2) application of the second N dose immediately before rather than after panicle initiation (De Datta *et al.*, 1987b).

In recent collaborative IRRI/CSIRO field studies, basal incorporation of urea with no standing water rather than broadcast application of urea into standing floodwater at 10 DT substantially reduced ammonia loss determined by the simplified bulk aerodynamic technique (Table 2). Total N loss, determined by the ^{15}N balance technique, was also reduced but not eliminated.

The effective reduction in gaseous N loss with basal incorporation of urea required application into drained soil. Basal incorporation of urea with 0.05-m standing water rather than no standing water resulted in nearly the same ammonia loss and total N loss as broadcast application into 0.05-m standing water at 10 DT (Table 2). Collaborative IRRI/CSIRO research at two additional sites in the Philippines consistently revealed no significant benefit of basal incorporation of urea with 0.05-m standing water over broadcast application of urea at 10 DT in reducing total gaseous N loss (Table 5).

Although basal incorporation of urea into drained soil can reduce N loss and frequently increase yield (De Datta *et al.*, 1987b), gaseous N loss remains substantial, suggesting that considerable scope for improving the efficiency of urea still exists. Implements and technologies for more thorough incorporation of prilled and granular urea merit investigation.

In Java, some farmers follow a practice known as *macak macak* in which water is drained from fields prior to broadcast application of prilled urea. Water is not returned until several days later. This practice is believed to be an effective management practice to prevent runoff loss of applied N in continuous flow irrigation systems (Wetselaar *et al.*, 1986). However, in the absence of runoff loss,

Table 5. Effect of method of urea application on total N loss from 80 kg urea-N/ha applied with 0.05-m standing floodwater in the Philippines

Site	Urea application		Total N loss[a]
	Timing	Method	
			% of applied N
Mabitac	Basal	Incorporation	58
	10 DT[b]	Broadcast	59
Calauan	Basal	Incorporation	62
	10 DT	Broadcast	65
Aguilar	Basal	Incorporation	62
	10 DT	Broadcast	60

[a] Determined from unaccounted-for ^{15}N in ^{15}N balances.
[b] DT designates d after transplanting.
Source: Unpublished IRRI/CSIRO collaborative research.

broadcast application of urea at 10 to18 DT to drained soil followed by flooding three days later was only marginally effective (De Datta *et al.*, 1987b) or completely ineffective (Figure 2) in reducing total gaseous N loss compared to broadcast application into 0.05-m water. Moreover, it failed to increase grain yield (Buresh *et al.*, 1988d).

Immediate flooding following broadcast application of urea to drained soil might be essential to achieve reductions in gaseous N loss and increases in grain yield. Researchers in China (Chen *et al.*, 1987) reported that broadcast application of urea to drained, unsaturated soil followed by immediate flooding resulted in higher yield than conventional broadcasting into floodwater. The flooding apparently moved the fertilizer-N into the soil where it was less prone to ammonia loss (Humphreys *et al.*, 1988).

C. Rate of Nitrogen Application

Research results on the relationship between loss of applied N and N application rate are inconsistent. Sudjadi *et al.* (1987) reported that pNH_3 following broadcast application of urea in West Java, Indonesia, increased linearly with rate of applied N. Hence, they concluded that ammonia loss was directly proportional to rate of applied urea-N. In Malaysia, Bidin *et al.* (1987) found that the percentage loss of urea determined by the ^{15}N balance technique can increase substantially at higher application rates, but the effect was not consistent between two sites. At Bumbong Lima, total N losses increased from 35 to 72% of the applied N when urea application rate increased from 56 to 125 kg N/ha, but in a more productive rice area at Tanjong Karang, the proportion of N lost (40%) was not affected by increasing the rate from 40 to 80 kg N/ha.

In a summary of ^{15}N balance studies with transplanted lowland rice, Vlek and Byrnes (1986) found no consistent relationship between urea rate and total N

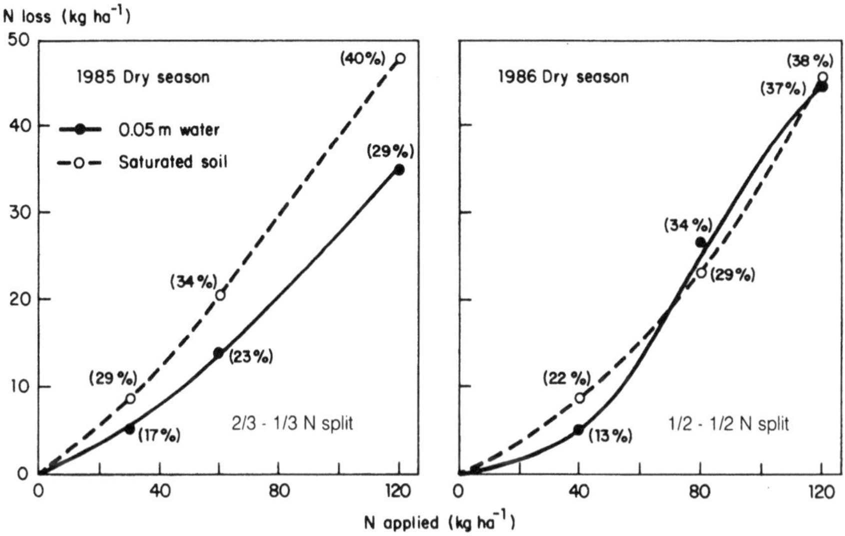

Figure 2. Effect of broadcasting urea (18 d after transplanting and 5 to 10 d after panicle initiation) to either 0.05-m standing water or saturated soil on N loss in the Philippines determined by the ^{15}N balance technique. The numbers in parentheses are percentages of applied N. (Adapted from Buresh *et al.*, 1988b and unpublished IFDC/IRRI collaborative research.)

loss. It is noteworthy that previous ^{15}N balance studies (Vlek and Byrnes, 1986) and gaseous N loss studies (Table 1) have tended to be for one timing of N application at a rate between 38 and 90 kg N/ha. In recent collaborative IRRI/IFDC studies, the effect of N rate on N loss was examined at rates as low as 20 kg N/ha for a broadcast urea application at 18 DT (Buresh *et al.*, 1988b). Table 6 shows that in each of the 2 trials, the proportion of N loss increased with increasing N rate.

The N losses in Table 6 represent total unaccounted-for ^{15}N for both times of N application. These N losses are generally lower than losses reported for one N application at 10 to 20 DT (Tables 1, 2, and 3) because a high N loss at 10 to 20 DT would be compensated for by lower N loss for the N application at a later plant growth stage (Fillery *et al.*, 1986c).

The results in Table 6 suggest that when an organic fertilizer source substitutes for part of a chemical fertilizer, the proportion of N loss from the chemical fertilizer might be reduced because of lower application rates. Alternatively, the results suggest that farmers might reduce the loss of N fertilizer through use of additional split doses. Research is needed to examine these hypotheses.

Nitrogen losses are known to occur from rice plants (Wetselaar and Farquhar, 1980), but the importance of this loss mechanism under field conditions has been uncertain. Buresh *et al.* (1988a) conducted sequential ^{15}N balances following application of 40 kg urea-N/ha either basal or at 11 DT, and they found no

Table 6. Effect of N rate on the total N loss from urea applied to rice at one site in the Philippines

Year	N rate: 16 or 18 DT[a]	N rate: 50–10 DAPI[b]	N rate: Total	Total N loss[c]
	kg/ha	kg/ha	kg/ha	% of applied N
1985	20	10	30	17
	40	20	60	23
	80	40	120	29
1986	20	20	40	13
	40	40	80	34
	60	60	120	37

[a] DT designates d after transplanting.
[b] DAPI designates d after panicle initiation.
[c] Determined from unaccounted-for ^{15}N in ^{15}N balances.
Source: Adapted from Buresh et al., 1988b.

increase in N loss from 20 DT to maturity. Hence, they concluded that N loss from plants was negligible in their study. The available field research suggests that gaseous N loss from plants is probably significant only at high rates of N application (Wetselaar *et al.*, 1986; De Datta *et al.*, 1988).

D. Prospect for Increasing Nitrogen Use Efficiency

We conclude that improved timing and incorporation of urea offer considerable prospect for increasing agronomic efficiency by both reducing N loss and better matching N availability to plant requirements at critical growth stages. Where the gaseous N loss is principally ammonia volatilization, it is possible to substantially reduce the loss by improved water management and thorough incorporation of fertilizer N into drained soil immediately before transplanting (Table 2). The use of algicides (Bowmer and Muirhead, 1987), urease inhibitors, and coatings offers another possibility for reducing ammonia loss. Urease inhibitors might increase runoff loss of fertilizer-N in environments with water flow from rice fields. Although urease inhibitors and algicides can reduce ammonia loss, they presumably do not reduce denitrification loss. In fact, it remains to be proven that they do not enhance denitrification loss (Byrnes, 1988).

Adjustment in water depth when broadcasting urea after transplanting has failed to reduce total gaseous N loss (Wetselaar *et al.*, 1986; De Datta *et al.*, 1987b; Freney *et al.*, 1988). Of course, broadcast application of urea to saturated soil rather than to floodwater could effectively reduce or prevent runoff loss (Wetselaar, 1985). Application of urea to drained, unsaturated soil followed by immediate flooding to move fertilizer-N into the soil has increased yield (Chen *et al.*, 1987) and merits further investigation.

Broadcast seeding of germinated rice seeds onto wet puddled soil is becoming a more important stand establishment technique. In a recent ^{15}N balance study,

De Datta *et al.* (1988) reported that losses of N fertilizer are less with broadcast seeded than transplanted rice. De Datta *et al.* (1988) attributed the differences in N loss between the two stand-establishment techniques to greater plant density and initial plant growth with broadcast seeded than transplanted rice. Research is needed to examine whether broadcast seeding rather than transplanting offers potential to increase the agronomic efficiency of rice.

Improved incorporation or placement of urea and coatings of urea would offer considerable potential for increasing the yield response of rice to urea fertilization. We suggest that priorities should be placed on (1) development of labor- and cost-effective methods of fertilizer placement, (2) dissemination of available information on the most agronomically effective timings for conventional prilled and granular urea, and (3) development of a cost-effective and environmentally safe coated urea.

VI. Supplemental Sources of Nitrogen

There has been a rising concern that heavy use of chemical fertilizers to the exclusion of organic fertilizer may have possible long-term detrimental effects related to soil structure, crop productivity, and off-farm pollution. Green manure and other organic fertilizers have a number of apparent agronomic and environmental advantages over chemical fertilizer (Bouldin, 1988). However, the economic feasibility of widespread adoption and use of green manure in rice-based cropping systems is less clear, and there has been relatively little analysis of the economics of green manuring in the tropics (Rosegrant and Roumasset, 1988).

One reason for increased interest in green manure crops in recent years is import substitution. For example, India imports fertilizer worth $1 billion (US) (Abrol and Palaniappan, 1988). Integrated use of green manure with chemical fertilizers might offer an opportunity to reduce a portion of the current high requirements for chemical fertilizer. A large number of crops and plant species have been used for green manuring. Table 7 gives the partial list of the green manure crops grown in India with their average green matter yield.

In China, milk vetch is an important green manure crop, seeds of which are usually broadcast between rice rows during heading of the late rice crop in order to prolong the growth period (Liu, 1988). *Sesbania* is used as green manure in south China rice fields, mainly in Fujian, Guangdong, Jiangsu and Zhejiang provinces. *Sesbania* contains high amounts of N, P, K, and microelements.

Azolla is another biological source of nitrogen that is used in China and Vietnam. However, its widescale adoption has been restricted by its P requirement and insect and disease problems. According to Liu (1988), it is not difficult to produce 75 Mg/ha fresh *Azolla* biomass during the winter season in China, provided there is no freezing weather longer than 5 d.

Evaluation of *Sesbania* and *Azolla* should involve research over several years of cropping in order to demonstrate their long-term effect on soil productivity and yield stability. With that objective in mind, several long-term trials have been

Table 7. Some common leguminous green manure crops in India

Local name	Botanical name	Growing season	Average green matter yield
			Mg/ha
Sunnhemp	*Crotolaria juncea*	Wet	21
Dhaincha	*Sesbania aculeata*	Wet	20
Pillipesera	*Phaseolus trilobus*	Wet	18
Mung	*Phaseolus aureus*	Wet	8
Cowpea	*Vigna sinensis*	Wet	15
Guar	*Cyamopsis tetragonoloba*	Wet	20
Senji	*Melilotus alba*	Dry	29
Khesari	*Lathyrus sativus*	Dry	12
Berseem	*Trifolium alexandrinum*	Dry	16

Source: Adapted from Abrol and Palaniappan, 1988.

established in Asia to evaluate green manure crops and *Azolla*, either as a single N source or in combination with chemical N fertilizers. Table 8 shows recent research results from two long-term trials in the Philippines using single and combined sources of N. *Azolla* and rice straw compost could substitute for part of the USG at Victoria but not at IRRI, where USG was significantly superior to prilled urea. Integrated use of an organic source with prilled urea was clearly superior to sole use of the organic N source in the case of rice straw compost but not fresh *Azolla*.

Nagarajah (1988) showed that ammonium N is rapidly released from incorporated *Sesbania* and *Azolla* in the tropics. Nonetheless, if this release of ammonium-N from green manure N does not meet the N requirements of rice or match the N uptake pattern of rice, supplemental applications of chemical N fertilizer will be needed for high yields. Morris *et al.* (1986a, 1986b) observed that incorporation of mung bean, *Vigna radiata* (L.) Wilczek, or cowpea green manure prior to rice transplanting increased rice yield by an average 2 Mg/ha. Timely application of additional N as chemical fertilizer to rice further increased yield.

John (1987) found that incorporation of cowpea green manure or residue 15 d prior to transplanting had no effect on N losses from urea either incorporated prior to transplanting or broadcast at 15 DT, as determined by the ^{15}N balance technique. Biswas (1988) observed that ^{15}N losses from a combined application of *Sesbania* green manure and urea were slightly less than from either sole *Sesbania* or sole urea. On the basis of pNH_3 measurements, he attributed the reduced ^{15}N loss to a reduction in ammonia loss, possibly because of a reduction in floodwater pH owing to CO_2 production from added green manure. Biswas (1988), unlike John (1987), incorporated green manure immediately prior to transplanting.

The optimum integrated N management strategy would be influenced by: (1) availability of native soil N, (2) mineralization of the organic N source, (3)

Table 8. Effects of single and combined N fertilizer source on grain yield for the tenth successive rice crop at two sites during the dry season in the Philippines

N source	Grain yield IRRI	Grain yield Victoria
	Mg/ha	
N fertilizer N	2.8 h	3.2 e
Single sources[a]		
PU, best split	5.7 bcd	6.6 abc
USG	6.7 a	7.2 a
Fresh *Azolla*, soil incorporated	5.4 de	6.9 ab
Fresh rice straw, soil incorporated	4.4 g	5.4 d
Rice straw compost, soil incorporated	4.9 f	5.4 d
Fresh *Sesbania*, soil incorporated	4.8 fg	NT[b]
Combined sources[c]		
Fresh *Azolla* + PU, best split	5.8 bcd	6.1 bcd
Fresh rice straw + PU, best split	5.2 ef	6.0 cd
Rice straw compost + PU, best split	5.6 cde	6.5 abc
Fresh *Azolla* + USG	6.1 b	7.0 a
Fresh rice straw + USG	6.0 bc	6.4 abc
Rice straw compost + USG	5.6 cde	6.7 abc
CV (%)	6	9

[a] N applied at 116 kg N/ha^{-1}. For best split application, two-thirds urea-N was broadcast and incorporated at planting and one-third urea-N was topdressed at 5 to 7 d before panicle initiation, PU designates prilled urea and USG designates urea supergranules.
[b] NT designates not tested.
[c] 58 kg N/ha applied each as organic source and inorganic source.
Source: Unpublished data.

timing of the chemical N source, (4) losses of native and added N, and (5) interactive effects of the organic and chemical fertilizers on N availability and N loss. Green manure has been shown to increase the agronomic efficiency of inorganic N fertilizer (Bouldin, 1988), but more research is needed to understand the effects of organic N sources on crop utilization and losses of chemical N fertilizer.

VII. Knowledge Gaps

Considerable scope still exists for improving the agronomic efficiency of N fertilizer. Even though the recommended incorporation of N fertilizer into drained soil before transplanting reduces ammonia loss, total N loss remains high (Table 2). More work is needed to identify and promote alternative N management strategies that eliminate or, at least, reduce N loss.

Because the effectiveness of a N management practice can depend on season, soil type, climate, water management, rice variety, and cropping pattern, there will probably be no universal best N management practice. Thus, more effort is needed to identify target environments for particular improved N management strategies. Improvements in N management could involve alternative water management, modified urea sources, alternative N timing and rates, and integration of organic and inorganic N fertilizers. Integration of N fertilizers should consider the rate of N release from organic fertilizers and the possibility of N losses from organic N sources (Bouldin, 1988). Little is known about the magnitude and processes of N loss from organic fertilizers.

References

Abrol, I.P., and S.P. Palaniappan. 1988. Green manure crops in irrigated and rainfed lowland rice-based cropping systems in South Asia, pp. 71-82. In: *Sustainable Agriculture-Green Manure in Rice Farming*. International Rice Research Institute, Los Baños, Laguna, Philippines.

Barker, R., R.W. Herdt, and B. Rose. 1985. *The rice economy in Asia*. Resources for the Future, Washington, DC, 324 pp.

Bidin, A., Z. Malek, and J.R. Simpson. 1987. Nitrogen loss from fertilizer applied to rice fields in Malaysia. Paper presented at the ACIAR Workshop on Gaseous Nitrogen Loss from Fertilizer in Asian Cropping Systems, 23–25 March 1987, Nanjing, China.

Biswas, T.K. 1988. Nitrogen dynamics and nitrogen-15 balance in lowland rice as affected by green manure and urea application. PhD diss. Indian Agricultural Research Institute, New Delhi, India.

Blackmer, A.M. 1986. Potential yield response of corn to treatments that conserve fertilizer nitrogen in soils. *Agron. J.* 78:571–575.

Bouldin, D.R. 1988. Effect of green manure on soil organic matter content and nitrogen availability, pp. 151-163. In: *Sustainable Agriculture – Green Manure in Rice Farming*. International Rice Research Institute, Los Baños, Laguna, Philippines.

Bowmer, K.H., and W.A. Muirhead. 1987. Inhibition of algal photosynthesis to control pH and reduce ammonia volatilization from rice floodwater. *Fert. Res.* 13:13–29.

Broadbent, F.E., S.K. De Datta, and E.V. Laureles. 1987. Measurement of nitrogen utilization efficiency in rice genotypes. *Agron. J.* 79:786–791.

Buresh, R.J. 1987. Relative susceptibility of conventional and experimental nitrogen sources to ammonia loss from flooded rice fields. *Fert. Res.* 13:139–153.

Buresh, R.J., and E.R. Austin. 1988. Direct measurement of dinitrogen and nitrous oxide flux in flooded rice fields. *Soil Sci. Soc. Am. J.* 52:681–688.

Buresh, R.J., S.K. De Datta, E.G. Castillo, and M.I. Samson. 1988a. Denitrification in irrigated and rainfed tropical lowland rice fields. Abstract from National Soils Conference, 9–12 May 1988, Australian Society of Soil Science, Canberra, Australia.

Buresh, R.J., S.K. De Datta, J.L. Padilla, and T.T. Chua. 1988b. Potential of inhibitors for increasing response of lowland rice to urea fertilization. *Agron. J.* 80:947–952.

Buresh, R.J., S.K. De Datta, J.L. Padilla, and M.I. Samson. 1988c. Effect of two urease inhibitors on floodwater ammonia following urea application to lowland rice. *Soil Sci. Soc. Am. J.* 52:856–861.

Buresh, R.J., S.K. De Datta, J.L. Padilla, and M.I. Samson. 1988d. Field evaluation of two urease inhibitors with transplanted lowland rice. *Agron. J.* 80:763–768.

Byrnes, B.H. 1988. The degradation of the urease inhibitor phenyl phosphorodiamidate in soil systems and the performance of *N*-(*n*-butyl) thiophosphoric triamide in flooded rice culture. Ph.D. diss. Technical University of Munich. Munich, Germany.

Cai Gui-Xin, Zhu Zhao-Liang, A.C.F. Trevitt, J.R. Freney, and J.R. Simpson. 1986. Nitrogen loss from ammonium bicarbonate and urea fertilizers applied to flooded rice. *Fert. Res.* 10:203–215.

Castillo, E.G. 1985. Effect of phenyl phosphorodiamidate (PPD) urease inhibitor on floodwater characteristics and ^{15}N recovery of urea in lowland rice. MS thesis. University of the Philippines at Los Baños, Laguna, Philippines.

Chen Lizhi. 1988. Green manure cultivation and use for rice in China, pp. 63–70. In: *Sustainable Agriculture—Green Manure in Rice Farming.* International Rice Research Institute, Los Baños, Laguna, Philippines.

Chen Rongye, Zhang Jiancai, Guo Wangmo, and Chen Wei. 1987. N fertilizer (urea) top-dressed on unsaturated soil and deep-placed using reflooding water. *Int. Rice Res. Newsletter* 12(5):35–36.

Craswell, E.T., B.H. Byrnes, L.S. Holt, E.R. Austin, I.R.P. Fillery, and W.M. Strong. 1985. Nitrogen-15 determination of nonrandomly distributed dinitrogen in air. *Soil Sci. Soc. Am. J.* 49:664–668.

Craswell, E.T., and D.C. Godwin. 1984. The efficiency of nitrogen fertilizers applied to cereals in different climates. pp. 1–55. In: P.B. Tinker and A. Lauchli (eds.) *Advances in plant nutrition*, vol. 1. Praeger, NY.

Craswell, E.T., and P.L.G. Vlek. 1979. Fate of fertilizer nitrogen applied to wetland rice. pp. 175–192. In: *Nitrogen and rice.* International Rice Research Institute, Los Baños, Laguna, Philippines.

De Datta, S.K. 1981. *Principles and practices of rice production.* Wiley, NY, 618 pp.

De Datta, S.K. 1987. Nitrogen transformation processes in relation to improved cultural practices for lowland rice. *Plant Soil* 100:47–69.

De Datta, S.K., and F.E. Broadbent. 1988. Methodology for evaluating nitrogen utilization efficiency by rice genotypes. *Agron. J.* 80:793–798.

De Datta, S.K., R.J. Buresh, W.N. Obcemea, and J.G. Real. 1987a. Direct field measurement of ammonia volatilization and denitrification in lowland rice soils. *Agron. Abstr.*, American Society of Agronomy, Madison, WI, p. 201.

De Datta, S.K., R.J. Buresh, M.I. Samson, and Wang Kai-Rong. 1988. Nitrogen use efficiency and nitrogen-15 balances in broadcast-seeded flooded and transplanted rice. *Soil Sci. Soc. Am. J.* 52:849–855.

De Datta, S.K., F.V. Garcia, A.K. Chatterjee, W.P. Abilay, Jr., J.M. Alcantara, B.S. Cia, and H.C. Jereza. 1979. "Biological constraints to farmers' rice yields in three Philippine provinces." IRRI Research Paper Series No. 30. International Rice Research Institute, Los Baños, Laguna, Philippines.

De Datta, S.K., and K.A. Gomez. 1981. Interpretive analysis of the international trials on nitrogen fertilizer efficiency in wetland rice. pp. 1–5. In: *Fertilizer International* (May issue). British Sulphur Corp., London, U.K.

De Datta, S.K., C.P. Magnaye, and J.C. Moomaw. 1968. Efficiency of fertilizer nitrogen (^{15}N-labeled) for flooded rice. *9th Int. Congr. Soil Sci. Trans.* (Adelaide, Australia) 4:67–76.

De Datta, S.K., W.N. Obcemea, R.Y. Chen, J.C. Calabio, and R.C. Evangelista. 1987b. Effect of water depth on nitrogen use efficiency and nitrogen-15 balance in lowland rice. *Agron. J.* 79:210–216.

De Datta, S.K., and W.H. Patrick, Jr. (eds.). 1986. *Nitrogen economy of flooded rice soils*. Martinus Nijhoff, Dordrecht, The Netherlands, 186 pp.

Denmead, O.T. 1983. Micrometeorological methods for measuring gaseous loss of nitrogen in the field. pp. 131–157. In: J.R. Freney and J.R. Simpson (eds.) *Gaseous loss of nitrogen from plant-soil systems*. Martinus Nijhoff/Dr. W. Junk Publishers, The Hague, The Netherlands.

FAO. 1987. *1986 FAO fertilizer yearbook*, vol. 36. Food and Agriculture Organization of the United Nations, Rome, Italy.

FAO. 1974. 1973 *Annual fertilizer review*, Food and Agriculture Organization of the United Nations, Rome, Italy.

Fillery, I.R.P., and B.H. Byrnes. 1984. Concurrent measurement of ammonia loss and denitrification in flooded rice fields. *Agron. Abstr.*, American Society of Agronomy, Madison, WI, p. 204.

Fillery, I.R.P., and S.K. De Datta. 1986. Ammonia volatilization from nitrogen sources applied to rice fields: I. Methodology, ammonia fluxes, and nitrogen-15 loss. *Soil Sci. Soc. Am. J.* 50:80–86.

Fillery, I.R.P., S.K. De Datta, and E.T. Craswell. 1986a. Effect of phenyl phosphorodiamidate on the fate of urea applied to wetland rice fields. *Fert. Res.* 9:251–263.

Fillery, I.R.P., P.A. Roger, and S.K. De Datta. 1986b. Ammonia volatilization from nitrogen sources applied to rice fields: II. Floodwater properties and submerged photosynthetic biomass. *Soil Sci. Soc. Am. J.* 50:86–91.

Fillery, I.R.P., J.R. Simpson, and S.K. De Datta. 1984. Influence of field environment and fertilizer management on ammonia loss from flooded rice. *Soil Sci. Soc. Am. J.* 48:914–920.

Fillery, I.R.P., J.R. Simpson, and S.K. De Datta. 1986c. Contribution of ammonia volatilization to total nitrogen loss after applications of urea to wetland rice fields. *Fert. Res.* 8:193–202.

Fillery, I.R.P., and P.L.G. Vlek. 1986. Reappraisal of significance of ammonia volatilization as an N loss mechanism in flooded rice fields. pp. 79–98. In: S.K. De Datta and W.H. Patrick, Jr. (eds.) *Nitrogen economy of flooded rice soils*. Martinus Nijhoff, Dordrecht, The Netherlands.

Freney, J.R., O.T. Denmead, I. Watanabe, and E.T. Craswell. 1981. Ammonia and nitrous oxide losses following applications of ammonium sulfate to flooded rice. *Aust. J. Agric. Res.* 32:37–45.

Freney, J.R., R. Leuning, J.R. Simpson, O.T. Denmead, and W.A. Muirhead. 1985. Estimating ammonia volatilization from flooded rice fields by simplified techniques. *Soil Sci. Soc. Am. J.* 49:1049–1054.

Freney, J.R., J.R. Simpson, and O.T. Denmead. 1983. Volatilization of ammonia. pp. 1–32. In: J.R. Freney and J.R. Simpson (eds.) *Gaseous loss of nitrogen from plant-soil systems*. Martinus Nijhoff/Dr. W. Junk Publishers, The Hague, The Netherlands.

Freney, J.R., A.C.F. Trevitt, W.A. Muirhead, O.T. Denmead, J.R. Simpson, and W.N. Obcemea. 1988. Effect of water depth on ammonia loss from lowland rice. *Fert. Res.* 16:97–107.

Hauck, R.D. 1986. Field measurement of denitrification—an overview. pp. 59–72. In: R.D. Hauck and R.W. Weaver (eds.) "Field measurement of dinitrogen fixation and denitrification." Spec. Pub. 18. SSSA, Madison, WI.

Hauck, R.D., S.W. Melsted, and P.E. Yankwich. 1958. Use of N-isotope distribution in nitrogen gas in the study of denitrification. *Soil Sci.* 86:287–291.

Heese, P.R. 1984. Potential of organic materials for soil improvement. pp. 35–43. In:

Organic matter and rice. International Rice Research Institute, Los Baños, Laguna, Philippines.

Holt, L.S., C.B. Christianson, E.R. Austin, and J.C. Katyal. 1988. A laboratory technique for releasing and measuring denitrification products trapped in soil. *Soil Sci. Soc. Am. J.* 52:1510–1511.

Humphreys, E., J.R. Freney, W.A. Muirhead, O.T. Denmead, J.R. Simpson, R. Leuning, A.C.F. Trevitt, W.N. Obcemea, R. Wetselaar, and Cai Gui-Xin. 1988. Loss of ammonia after application of urea at different times to dry-seeded, irrigated rice. *Fert. Res.* 16:47–57.

IFDC. 1988. Annual report for 1987. International Fertilizer Development Center, Muscle Shoals, AL.

Iimura, K., and J. Ito. 1984. Direct measurement of N_2 denitrified from soil and fertilizer in paddy fields using ^{15}N as a tracer. *JARQ* 18:27–30.

IRRI. 1986a. "Annual report for 1985." International Rice Research Institute, Los Baños, Laguna, Philippines.

IRRI. 1986b. "World rice statistics 1985." International Rice Research Institute, Los Baños, Laguna, Philippines.

IRRI. 1987. "Annual report for 1986." International Rice Research Institute, Los Baños, Laguna, Philippines.

IRRI. 1988. "Annual report for 1987." International Rice Research Institute, Los Baños, Laguna, Philippines.

Ishikawa, M. 1988. Green manuring in rice: The Japan experience. p. 45–61. In: *Sustainable Agriculture – Green manure in rice farming*. International Rice Research Institute, Los Baños, Laguna, Philippines.

John, P.S. 1987. Nitrogen economy in rice based cropping systems through cowpea green manure, crop residue, and fertilizer nitrogen management. PhD diss. Indian Agricultural Research Institute, New Delhi, India.

Katyal, J.C., L.S. Holt, and A.M. Gadalla. 1989. A method to determine soil-entrapped denitrification products in submerged soils. *Soil Sci. Soc. Am. J.* 53: (In press).

Katyal, J.C., Bijay Singh, P.L.G. Vlek, and E.T. Craswell. 1985. Fate and efficiency of nitrogen fertilizers applied to wetland rice II. Punjab, India. *Fert. Res.* 6:279–290.

Keeney, D.R. 1986. Critique of the acetylene blockage technique for field measurement of denitrification. pp. 103–115. In: R.D. Hauck and R.W. Weaver (eds.) "Field measurement of dinitrogen fixation and denitrification." Spec. Pub. 18. SSSA, Madison, WI.

Leuning, R., O.T. Denmead, J.R. Simpson, and J.R. Freney. 1984. Processes of ammonia loss from shallow floodwater. *Atmos. Environ.* 18:1583–1592.

Leuning, R., J.R. Freney, O.T. Denmead, and J.R. Simpson. 1985. A sampler for measuring atmospheric ammonia flux. *Atmos. Environ.* 19:1117–1124.

Lindau, C.W., W.H. Patrick, Jr., R.D. DeLaune, K.R. Reddy, and P.K. Bollich. 1988. Entrapment of nitrogen-15 dinitrogen during soil denitrification. *Soil Sci. Soc. Am. J.* 52:538–540.

Liu Chungchu. 1988. Integrated use of green manure in ricefields in South China. pp. 319–331. In: *Sustainable Agriculture – Green Manure in Rice Farming*. International Rice Research Institute, Los Baños, Laguna, Philippines.

Maene, L.M., A. de Vuyst, and S.P. Pradhan. 1987. The importance of urea as a source of nitrogen for crop production in the Asian and Pacific region. Paper presented at the International Symposium on Urea Technology and Utilization, 16–19 March 1987, Kuala Lumpur, Malaysia.

McCune, D.L. 1982. Fertilizers for tropical and subtropical agriculture. 12th Francis New Memorial Lecture. Presented before the Fertilizer Society of London (12 March

1981). IFDC Spec. Pub. SP-2, International Fertilizer Development Center, Muscle Shoals, AL.
Mikkelsen, D.S., S.K. De Datta, and W.N. Obcemea. 1978. Ammonia volatilization losses from flooded rice soils. *Soil Sci. Soc. Am. J.* 42:725–730.
Moraghan, J.T., and R.J. Buresh. 1977. Correction for dissolved nitrous oxide in nitrogen studies. *Soil Sci. Soc. Am. J.* 41:1201–1202.
Morris, R.A., R.E. Furoc, and M.A. Dizon. 1986a. Rice responses to a short-duration green manure. I. Grain yield. *Agron. J.* 78:409–412.
Morris, R.A., R.E. Furoc, and M.A. Dizon. 1986b. Rice responses to a short-duration green manure. II. N recovery and utilization. *Agron. J.* 78:413–416.
Mosier, A.R., W.D. Guenzi, and E.E. Schweizer. 1986. Soil losses of dinitrogen and nitrous oxide from irrigated crops of northeastern Colorado. *Soil Sci. Soc. Am. J.* 50:344–348.
Nagarajah, S. 1988. Transformation of green manure nitrogen in lowland rice soils. pp. 193–208. In: *Sustainable Agriculture – Green Manure in Rice Farming*. International Rice Research Institute, Los Baños, Laguna, Philippines.
Parish, D.H., L.L. Hammond, and E.T. Craswell. 1980. Research on modified fertilizer materials for use in developing-country agriculture. IFDC Paper Series P-2. International Fertilizer Development Center, Muscle Shoals, AL.
Prasad, R., and S.K. De Datta. 1979. Increasing efficiency of fertilizer nitrogen in wetland rice. pp. 465–484. In: *Nitrogen and rice*. International Rice Research Institute, Los Baños, Laguna, Philippines.
Prins, W.H., G.D. van Brakel, and T. van der Sar. 1984. Pneumatic injector for deep placement of urea in wetland rice soils. Paper presented at 2d International Conference on the Development of the Agricultural Machinery Industry in Developing Countries, Amsterdam, The Netherlands.
Rao, D.L.N. 1987. Slow-release urea fertilizers – effect on floodwater chemistry, ammonia volatilization and rice growth in an alkali soil. *Fert. Res.* 13:209–221.
Reddy, K.R., W.H. Patrick, Jr., and C.W. Lindau. 1987. Fertilizer nitrogen losses in the root zone of rice. *Agron. Abstr.*, American Society of Agronomy, Madison, WI, p. 191.
Rolston, D.E., D.L. Hoffman, and D.W. Toy. 1978. Field measurement of denitrification: I. Flux of N_2 and N_2O. *Soil Sci. Soc. Am. J.* 42:863–869.
Rosegrant, M.W., and J.A. Roumasset. 1988. Economic feasibility of green manure in rice-based cropping systems. pp. 11–27. In: *Sustainable Agriculture – Green Manure in Rice Farming*. International Rice Research Institute, Los Baños, Laguna, Philippines.
Savant, N.K. 1988. How to use the IFDC dispenser method of hand deep placement of urea supergranules in transplanted rice. International Fertilizer Development Center, Muscle Shoals, AL.
Savant, N.K., and S.K. De Datta. 1982. Nitrogen transformations in wetland rice soils. *Adv. Agron.* 35:241–302.
Savant, N.K., and P.J. Stangel. 1985. Suggestions for efficient use of urea supergranules for wetland rice. pp. 133–142. In: P.J. Stangel *et al.* (eds.) Proceedings of the workshop on urea deep placement technology. IFDC Spec. Pub. SP-6. International Fertilizer Development Center, Muscle Shoals, AL.
Schnier, H.F., S.K. De Datta, K. Mengel, E.P. Marqueses, and J.E. Faronilo. 1988. Nitrogen use efficiency, floodwater properties, and nitrogen-15 balance in transplanted lowland rice as affected by liquid urea band placement. *Fert. Res.* 16:241–255.
Seiler, W., A. Holzapfel-Pschorn, R. Conrad, and D. Scharffe. 1984. Methane emission from rice paddies. *J. Atmos. Chem.* 1:241–268.
Siegel, R.S., R.D. Hauck, and L.T. Kurtz. 1982. Determination of $^{30}N_2$ and applica-

tion to measurement of N_2 evolution during denitrification. *Soil Sci. Soc. Am. J.* 46:68–74.

Simpson, J.R., and J.R. Freney. 1986. Interacting processes of gaseous nitrogen loss from urea applied to flooded rice fields. 13th Int. Congr. Soil Sci. Trans. (Hamburg, Germany) 3:968–969.

Simpson, J.R., J.R. Freney, R. Wetselaar, W.A. Muirhead, R. Leuning, and O.T. Denmead. 1984. Transformations and losses of urea nitrogen after application to flooded rice. *Aust. J. Agric. Res.* 35:189–200.

Smith, C.J., M. Brandon, and W.H. Patrick, Jr. 1982. Nitrous oxide emission following urea-N fertilization of wetland rice. *Soil Sci. Plant Nutr.* 28:161–171.

Stangel, P.J. 1985. Changes in the form and price of fertilizer—a global perspective. ASPAC Ext. Bull, 217. Food and Fertilizer Technology Center, Taiwan, Republic of China.

Stangel, P.J., and S.K. De Datta. 1985. Availability of inorganic fertilizers and their management – a focus on Asia. Paper presented at the International Rice Research Conference, 1–5 June 1985, International Rice Research Institute, Los Baños, Laguna, Philippines.

Strong, W.M., E.R. Austin, L.S. Holt, and R.J. Buresh. 1987. Determination of the combined nitrogen-15 content of dinitrogen and nitrous oxide in air. *Soil Sci. Soc. Am. J.* 51: 1344–1350.

Sudjadi, M., Y. Prawirasumantri, and R. Wetselaar. 1987. Nitrogen fertilizer efficiency in lowland rice in Indonesia, pp. 123–134. In: *Efficiency of nitrogen fertilizers for rice.* International Rice Research Institute, Los Baños, Laguna, Philippines.

Tang, A.M., and B. Stone. 1980. Food production in the People's Republic of China. IFPRI Res. Rep. 15. International Food Policy Research Institute, Washington, DC.

Vlek, P.L.G., and B.H. Byrnes. 1986. The efficacy and loss of fertilizer N in lowland rice, pp. 131–147. In: S.K. De Datta and W.H. Patrick, Jr. (eds.) *Nitrogen economy of flooded rice soils.* Martinus Nijhoff, Dordrecht, The Netherlands.

Vlek, P.L.G., and E.T. Craswell. 1981. Ammonia volatilization from flooded soils. *Fert. Res.* 2:227–245.

Watanabe, I., and M.R. de Guzman. 1980. Effect of nitrate on acetylene disappearance from anaerobic soil. *Soil Biol. Biochem.* 12:193–194.

Watanabe, I., B.C. Padre, Jr., and S.T. Santiago. 1981. Quantitative study on nitrification in flooded rice soil. *Soil Sci. Plant Nutr.* 27:373–382.

Wetselaar, R. 1985. Deep point-placed urea in a flooded soil: A mechanistic view. pp. 7–14. In: P.J. Stangel *et al.* (eds.) Proceedings of the workshop on urea deep placement technology. IFDC Spec. Pub. SP-6. International Fertilizer Development Center, Muscle Shoals, AL.

Wetselaar, R., and G.D. Farquhar. 1980. Losses of nitrogen from tops of plants. *Adv. Agron.* 33:263–302.

Wetselaar, R., N. Sri Mulyani, Hadiwahyono, A.M. Damdam, and S. Moersidi. 1986. The fate of fertilizer nitrogen in irrigated lowland rice in S.E. Asia: Water and plant influences and their practical implications. 13th Int. Congr. Soil Sci. Trans. (Hamburg, Germany) 4:772–782.

Wilson, J.D., V.R. Catchpoole, O.T. Denmead, and G.W. Thurtell. 1983. Verification of a simple micrometeorological method for estimating the rate of gaseous mass transfer from the ground to the atmosphere. *Agric. Meteorol.* 29:183–189.

Wilson, J.D., G.W. Thurtell, G.E. Kidd, and E.G. Beauchamp. 1982. Estimation of the rate of gaseous mass transfer from a surface source plot to the atmosphere. *Atmos. Environ.* 16:1861–1867.

Tropical Soils and Soil Classification Updates

A. Van Wambeke*

I. Introduction

Several national and international institutions are updating the soil classification systems they use in their soil resource inventory programs. The Food and Agriculture Organization (FAO) of the United Nations (1988), the Soil Survey Staff (1987) of the United States Department of Agriculture (USDA) Soil Conservation Service (SCS), several institutions in France and French-speaking African countries, and EMBRAPA (Empresa Brasileira de Pesquisa Agropecuária) in Brazil, among others, are revising their systems.

*Department of Agronomy, Cornell University, Ithaca, NY 14853 USA

Advances in Soil Science, Volume 10

This chapter reviews recent trends in the classification of soils of the tropics as they appear in taxonomic updates, particularly the "Keys to Soil Taxonomy" by the Soil Survey Staff (1987), and the revised legend of the *FAO-UNESCO Soil Map of the World* (FAO, 1988). Revisions by other institutions that are not published yet are not included in this review.

The chapter concludes with a discussion of soil classification systems that emphasizes their international dimensions and their contributions to agricultural sciences.

II. Keys to Soil Taxonomy 1987

The new version of "Keys to Soil Taxonomy" (Soil Survey Staff, 1987) contains several new features that imply changes to the original concepts, approaches, and intents that were standard at the time the first edition of *Soil Taxonomy* (Soil Survey Staff, 1975) was published. Although the new keys may not seem markedly different from previous printings, many amendments involve basic principles of soil classification.

The most important changes are the introduction of nonmutually exclusive subsurface diagnostic horizons, for example, the kandic horizon versus the argillic horizon; the new definition and classification of Oxisols; and a new approach to subdivide great groups into subgroups.

Most changes are the result of recommendations made by several international committees sponsored by the Soil Management Support Services (SMSS) of the Soil Conservation Service. The committees are organized by soil orders. The changes are implemented as official amendments only in those soil orders for which the committee has submitted a final report. The other taxa are left untouched. Alfisols, Ultisols, and Oxisols have been amended. The other orders will be revised later. The new keys therefore seem heterogeneous because some changes now introduced in some orders are not paralleled in others where they may be equally justified. When studying *Soil Taxonomy*, teachers and students should be aware that institutional structures are at the origin of the lack of uniformity across the soil orders.

A. Changes to Diagnostic Subsurface Horizons

1. The Kandic Horizon

The kandic horizon is a new diagnostic subsurface horizon. The original intent for its introduction at the time it was proposed (Moormann, 1975) was to upgrade the oxic subgroups of tropical Ultisols and Alfisols to the great soil group level. Oxic subgroups were mainly recognized on the basis of the NH_4OAc, pH 7 cation exchange capacity (CEC), which had to be less than 24 cmol(+) kg^{-1} clay. This upgrading would give due recognition to the kaolinitic mineralogy of many

Alfisols and Ultisols of tropical regions, which separates them clearly from most Alfisols and some Ultisols of temperate, generally glaciated areas.

The kandic horizon was also introduced to overcome serious difficulties in the diagnosis of argillic horizons in kaolinitic materials. These difficulties made reproducible identification of soils in many instances almost impossible.

The kandic horizon differs in its conception from other diagnostic subsurface horizons used to distinguish *mature* soil orders. It is not considered an expression of one *single* soil-forming process. Instead, it is recognized that *several, often unrelated* processes may be responsible for its formation. The clay increase in the kandic horizon, for example, may be the result of eluviation-illuviation, clay destruction, selective erosion, sedimentation, or lithological discontinuities.

The kandic horizon differs in other ways from other subsurface diagnostic horizons used at the order level. It is not a horizon that is mutually exclusive with all others used at the same level; for example, it is not mutually exclusive with the argillic horizon, nor is one of them a subdivision of the other: the same portion of a pedon may at the same time meet the requirements of an argillic and a kandic horizon and qualify for both. The two definitions overlap.

a. Relation with the Argillic Horizon

The kandic horizon shares important attributes both with the *argillic and the oxic* horizons. With respect to the *argillic*, it has a significant clay increase at its upper boundary, *almost but not completely* identical to the increases in clay required for the argillic horizon, and over a shorter distance (less than 15 cm instead of 30 cm). The clay increase thus associates the kandic with the argillic, but no argillans are required for the kandic horizon. The clay increase necessary to identify a kandic horizon will be called a *sharp* clay increase herein.

b. Relation with the Oxic Horizon

With respect to the oxic, the two horizons are similar in the nature of their clay fractions, which in both cases are low-activity clays. Low-activity clays are defined here as having an apparent CEC of less than 16 cmol(+) kg^{-1} clay by NH_4OAc at pH 7 or as having an apparent effective CEC of less than 12 cmol(+) kg^{-1} clay. Unlike the oxic horizon, however, the kandic horizon *may* contain appreciable amounts of weatherable minerals in the sand and silt fractions.

The kandic and the oxic horizons are mutually exclusive. They exclude themselves by the *sharp* clay increase, which is not allowed in oxic horizons, but is required in all kandic horizons.

c. Use of the Kandic Horizon in the Key

In the identification of Oxisols, the kandic horizon is used twice at the same level in the key: once to *exclude* soils from the Oxisols (when the surface layer contains less than 40% clay); once to *include* soils in the Oxisols (when the surface horizon contains at least 40% clay). The use of one diagnostic horizon in two opposite directions at the same point in the hierarchical structure causes difficulties for teachers and students in understanding the rationale followed to define orders.

2. The Oxic Horizon

There have been changes in the definition of the *oxic horizon*. In previous versions the argillic and the oxic horizon were mutually exclusive and the argillic had precedence over the oxic horizon. Oxisols, with some exceptions, could not have argillic horizons. This restriction has been removed in the new keys, and a given horizon may at the same time be argillic and oxic. Examples are horizons where the required clay increases at the upper boundary take place over a distance of more than 15 cm but less than 30 cm or horizons that qualify for argillic on the basis of argillans alone, as is often the case under plow layers.

In the key to the soil orders, the oxic criterion comes before the argillic; consequently, the oxic horizon now has precedence over the argillic. The Oxisol concept has thus been broadened considerably to include soils with argillic horizons.

This important change in the definition of the oxic horizon eliminates the difficulties of the past to identify argillic horizons in materials dominated by low-activity clays. It will be welcomed by many soil surveyors who work in tropical regions and who previously had constant problems with the identification of clay cutans and argillic horizons. Many European soil scientists considered that the earlier edition of *Soil Taxonomy* (Soil Survey Staff, 1975) had given too much diagnostic weight to the argillic horizon in materials that are strongly weathered.

3. The Use of Nonmutually Exclusive Diagnostic Horizons

There is, in principle, no major argument against the use of nonmutually exclusive diagnostic horizons. If users are aware that the new keys use subsurface diagnostic horizons with overlapping definitions, they can prepare in their identification procedures for such possibilities and check which horizon is critical at a given level in the key.

In previous editions, the spodic, argillic, oxic and cambic horizons were mutually exclusive, and one portion of a profile could only belong to one of them. This made the methodology to classify soils straightforward: the surveyor first identified the diagnostic horizon (if there was one, there was no worry about the others), then went to the key to determine the soil order. The present approach is different, is less straightforward, and may require more time for teaching it.

B. Changes to Taxa

1. Oxisols

a. Definition

The new definition of Oxisols results in a widening of the concept. Some soils, mostly clayey in the surface horizons, that were previously excluded because of the presence of an argillic horizon are now accepted.

The restriction on clay content has also been removed: the minimum requirement of 15% clay in the oxic horizon has been replaced by a texture requirement that excludes sands and loamy sands. It is not known whether this change will widen the Oxisol concept because the calculation of the *apparent* CEC or ECEC (effective cation exchange capacity) of the clay, without correction for organic matter, often results in values that exceed the critical limits for oxic horizons in materials with little clay.

b. The Identification of Oxisols

The criteria that are used to identify Oxisols in the new keys differ according to the texture of the surface horizon and the presence or absence of the *sharp* clay increase.

For any surface texture, when no *sharp* clay increase is present (*and consequently no kandic horizon occurs*), the oxic horizon is sufficient to identify Oxisols. As stated before, an argillic horizon may be present.

However, if the clay content is 40% or more in the surface 18 cm, both the kandic horizon, *with its sharp textural upper boundary*, or the oxic horizon can be used to *include* pedons in the Oxisols. In this case, the kandic horizon should not have more weatherable minerals than is allowed in an oxic horizon.

It is difficult to explain the rationale for using the same criterion, the *sharp* clay increase, in two opposite directions at the same place in the key: once to include soils in the Oxisols and once to exclude soils from the Oxisols. The way the kandic horizon is used at this level departs from the principle adhered to by *Soil Taxonomy* (Soil Survey Staff, 1975) that soils of one order should be the result of a unique set of pedogenic processes.

In the 1987 keys leading to Oxisols, the 40% clay-content criterion in the topsoil overrules the kandic horizon in low-activity clay materials. The developers of the new keys considered that in this way the most satisfactory grouping was obtained for the pedons examined by the international committees. In their views, soils with less than 40% clay in the topsoil and the *sharp* clay increase should either be Alfisols or Ultisols, regardless of their weathering stage, and not be associated with Oxisols. This view is probably influenced by the fact that in the United States the concept of the "modal" Oxisol originated from observations of soils formed in basic and ultrabasic materials in which horizons usually have very diffuse boundaries.

c. Subdivisions

Suborders. At the suborder level, soil moisture regimes are used to subdivide the Oxisols. For the first time, an atmospheric climate is used to distinguish a suborder: the perudic regime, which is based only on precipitation and potential evapotranspiration data, and characterizes the Perox suborder.

The name *Orthox* has been replaced by *Udox*. The formative element *orth*-, "the common one," seemed appropriate, and no valid explanation is given for the change. The broadened Oxisol concept now may cover soils with xeric soil moisture regimes ("Xerox"?) that key out in the Udox.

The Humox have been deleted and redistributed as *humic* subgroups.

Great Groups. Parallel great groups have been established uniformly in the Udox, Ustox, and Perox suborders. The keys for the great groups in each of these suborders consider the following sequence of diagnostic features: the sombric horizon, acric properties, high base saturation, and the kandic horizon. This results for example in the Ustox in a sequence of Sombriustox, Acrustox, Eutrustox, Kandiustox, and Haplustox.

Subgroups. Some new criteria have been introduced at the subgroup level that were not used in the 1975 *Soil Taxonomy*. Color, the kandic horizon, and organic matter content are used to distinguish, for example, xanthic, rhodic, kandiudalfic and humic subgroups.

There are no ultic or tropeptic subgroups anymore. The deletion of the ultic subgroup seems surprising because some soils with argillic horizons have now been accepted in the Oxisols.

The humic subgroups in the sombri- great groups seem geared in the wrong direction. Most Oxisols with sombric horizons occur at high elevation in the cool tropics and normally have "humic" properties; those that do not have "humic" properties are usually eroded, and are now called *typic*. If *typic* is used to designate the central concept of a great group, then no "humic" intergrades are necessary in the sombri- great groups. If, on the contrary, central concepts are not part of the approach followed by the new subgroup subdivisions in Oxisols, it would have been more appropriate to call the last subgroup in the key *haplic* instead of *typic*.

Considerable changes have been made at the subgroup level in Oxisols. The 1983 printing "Keys to Soil Taxonomy" recognized 23 subgroups; the 1987 version increased that number to 215, an almost tenfold increase. There are several reasons at the origin of this proliferation of subgroups.

a. During the development of *Soil Taxonomy* (1975), which went through a series of approximations, no taxon was *established* unless the actual occurrence of soils that would become members of the new taxon had been verified and their importance evaluated by mapping. There was no such thing as a *theoretical* taxon *created* to accomodate soils that still had to be discovered. This principle is not followed anymore in the 1987 keys, and many subgroups of Oxisols have no field representation.

b. The new structure of the subdivisions in the subgroup category of Oxisols, which abandoned the initial definition of a *typic* subgroup followed by a series of intergrades and extragrades is, in fact, a closed key, where the last taxon in the subgroup key accepts all pedons that have not been identified by previous sections in the key. The subgroup level thus becomes comprehensive for all existing and nonexisting theoretical pedons. The new structure eliminates implied subgroups that generated considerable soil correlation work. The new approach provides in advance niches to accommodate soils that the International Committee on Oxisols (ICOMOX) thinks are most likely to exist.

c. Some suborders have been eliminated (e.g., the Humox) and the soils that belonged to them have been spread as subgroups in several great groups where they are now recognized on the basis of *one and the same* attribute across the whole order. Forty subgroups now replace the Humox. The objective of a classification is to *group* soils that share the same property—indicative of common pedogenic processes—and that have a geographic expression in the field. A classification is not well served by separating like soils and redistributing them as small taxa at lower levels.

2. Ultisols and Alfisols

a. Definitions

The definitions of the two orders that previously made the presence of an argillic horizon mandatory have been changed by the inclusion of the kandic horizon as one of the criteria that opens the door to these orders.

Strongly weathered soils with low-activity clay mineralogy and a sharp clay increase are now accepted in these orders, even if they do not have an argillic horizon. The new definitions improve the effectiveness of the system by eliminating the difficulties in recognizing argillic horizons in low-activity clay soils. The grouping of "kandic" soils with the Ultisols and Alfisols accepts, in part, the hypothesis that most kandics are old argillics that have now reached the ultimate weathering stage. The developers of the new keys for some reason, however, only opened the door of the Ultisols and Alfisols to pedons whose surface layers contained less than 40% clay.

b. Subdivisions

Suborders. There are no changes at the suborder level, and contrary to the elimination of Humox in the Oxisols, the Humults have been maintained in the Ultisols.

Great Groups. Kandi- and Kanhapl- great groups have been introduced in the Humults, Aquults, Udults, Ustults, Aqualfs, Udalfs, and Ustalfs. They have not been introduced in the Xerults or Xeralfs.

Part of the original intent to upgrade the oxic subgroups of Alfisols and Ultisols to the great group level has thus been achieved. The amendments, however, did not pick up all the previous "oxic" subgroups because the critical CEC limit of 16 was used instead of 24.

In the key to great groups the *plinthic-* criterion (presence of at least 50% plinthite at less than 150-cm depth) comes first; for that reason, none of the kandi- or kanhapl- great groups can have plinthite in the amounts and at the depths mentioned.

The kandic and kanhaplic great groups replace the previous Trop- great groups in Udults and Aquults, which were defined on the basis of soil temperature regimes only. The kandic horizon thus achieves a marked improvement in the classification by using morphologic features that can be identified at all times instead of temperature criteria that need repeated and long-term measurements in the field to take into account seasonal changes. Trop- suborders and great groups are maintained in the Entisols, Inceptisols, Histosols, and Spodosols, however.

The inclusion of the low-activity clay soils with kandic horizons in the Alfisols and Ultisols does not remove the difficulties in the measurement of base saturation in soils with very low CEC and ECEC. Base saturation continues to be used to separate Ultisols from Alfisols although determinate errors on this property may reach 30% absolute for high-precision analytical methods in samples having a CEC of 10 cmol(+) kg^{-1} soil and a 50% base saturation. This point will be further discussed later.

Subgroups. There are still some subgroups defined by a CEC of less than 24 cmol(+) kg^{-1} clay in the great groups of Haplustalfs, Paleustalfs, Rhodustalfs, Plinthaquults, and Haplustults. Their definition corresponds to the old definitions of the "oxic" subgroups of previous versions of *Soil Taxonomy.* These subgroups include the soils that the "kandic" amendment did not upgrade to the great group level. They are variously called *kandic* or *kanhaplic* subgroups.

C. Significance of Changes

1. Illuviation Versus Weathering

The changes that are introduced in the new "Keys to Soil Taxonomy" are responses to demands from pedologists from several parts of the world with different views and experiences, and it cannot be expected that the amendments will satisfy the expectations of all of them.

The problem that the International Committee on Low Activity Clay Soils (ICOMLAC) addressed in the beginning of its activities, that is, to upgrade the "oxic" subgroups of tropical areas to the great group level, has been partly solved. This will be accepted by most pedologists with experience in the tropics as an improvement.

The need to upgrade the subgroups is actually the result of the strong taxonomic weight that *Soil Taxonomy* (1975) gives to the argillic horizon. Most soil scientists of temperate regions consider the argillic horizon to be the master horizon of mature land surfaces and therefore prefer to keep it at the highest level in the hierarchy of the system. In *Soil Taxonomy,* the argillic horizon only comes second in the Mollisols and Aridisols. Until the publication of the new keys, the argillic horizon excluded soils from the Oxisols. This restriction has been removed and replaced by the *sharp* clay increase for soils with less than 40% clay in the surface 18 cm.

Many soil scientists who have worked in Africa consider that extreme weathering imparts more important properties to soils than clay illuviation, and they argue that in low-activity clay soils, the argillic horizon is used above its ceiling of independence and separates like soils that belong together. Low-activity clay properties without restriction of clay increases should have the same power as the mollic epipedon and the aridic moisture regime to establish an order, just as in the case of Mollisols and Aridisols.

The new keys went halfway in that direction. For some, the amendments did not go far enough, and the upgrading should have gone to the order level. For others, who want to keep the Oxisol concept pure, the amendments went too far or the compromise lead to complicated pathways for which the rationale is difficult to follow.

2. Closing Soil Taxonomy?

The subgroup level of *Soil Taxonomy* (Soil Survey Staff, 1975) was the only noncomprehensive level in the system. It included only the soils that have been identified in the field; it provided room for new soils still to be discovered.

The new subdivisions of Oxisols, by using a key approach with a typic subgroup that catches at the end all the pedons not identified by previous clauses in the key, closes *Soil Taxonomy* at all levels. For some, this approach conflicts with the approximation concept of *Soil Taxonomy*, that is to reflect present knowledge and to provide open doors to incorporate new knowledge. With the present keys to the Oxisol subgroups and the creation of "theoretical" taxa with no representation in the field, it is felt that central concepts in soil classification are abandoned. For young soil scientists the new approach may be less attractive because the impression is given that there is nothing to discover anymore; at any rate, when using the new subgroup keys, classifiers do not receive a signal from the identification procedures that they may be dealing with a soil that had not been registered previously.

The reasons for adopting a closed key may be one of the following: (1) the acceptance that central concepts only exist in the minds of people and that the soil cover is a continuum without taxonomic clusters, (2) the fact that soil surveyors who have to produce maps need names for all soils and cannot make any progress with "unknowns," or (3) the avoidance of the proliferation of implied subgroups that are difficult to handle administratively.

3. Horizontal Versus Vertical Updating

The updates have been limited to certain orders down to the family level (vertical), without concern to modify parallel taxa in other orders (horizontal). Examples are the elimination of Humox, keeping Humults; the replacement of Trop-great groups in Ultisols, keeping Tropepts in Inceptisols; and the differences in the critical depths of plinthite in Plinthaquults (1.5 m) and Plinthaquox (1.25 m).

This unevenness reduces students' acceptance of the system as a scientific language. The recognition that natural systems contain separate kingdoms having their own class subdivisions in these cases does not change their views.

The updating procedures order by order are the result of the committee approach favored by SCS and SMSS. The system is slow. It took 12 years to implement in printing the ICOMLAC proposals. The ICOMOX amendments took almost a decade.

During the development of *Soil Taxonomy* before 1975, changes were usually included across orders and the classification as a whole changed to reach consistent application of accepted amendments in all taxa.

III. The FAO–Unesco 1988 Update

The legend of the FAO–UNESCO *Soil Map of the World* in its original version (FAO, 1974) included a two-level classification supported by an identification key. The classification has been revised by a group of soil scientists who introduced substantial changes to definitions of some major diagnostic horizons and added a third categorical level to the classification. The new legend is to serve soil maps of countries at the 1:1,000,000 scale.

The FAO classification has received wide acceptance in both the developing and developed countries; an example is the soil map of the European Economic Community (EEC) (Commission of European Communities, 1985).

The FAO–UNESCO 1974 legend used almost the same diagnostic properties as *Soil Taxonomy* (Soil Survey Staff, 1975); however, FAO–UNESCO combines these criteria in a different way. In spite of this, the two systems were easily comparable, although only a few taxa were identical. The changes now introduced by *Soil Taxonomy* and the FAO–UNESCO legend indicate that the parallelism between them is being broken and that they are drifting apart particularly in the case of tropical soils.

A. Structure of the FAO–UNESCO Legend

The structure of the FAO–UNESCO system will be discussed under these rubrics: (1) the number of highest level taxa (27 instead of 10 in *Soil Taxonomy*), (2) absence of a soil climatic component in the definition of the taxa, (3) the tentative approach to the third-level taxa.

1. Number of Highest Level Taxa

The presence of 27 classes at the highest categorical level increases the acceptability of the system. The reason for this is that when the soil universe is only divided into 10 classes, as in *Soil Taxonomy*, inevitably a number of contrasting soils are forced into one order where they do not seem to belong together. At the same time, when more classes are available, more soils find a place in one of the "mature" orders, and a smaller number of them end up in a catchall heterogeneous class at the bottom of the identification key. Finally, the variability within an FAO order is less than in one of the 10 *Soil Taxonomy* orders; consequently, an FAO highest level unit provides more precise information on soil properties than *Soil Taxonomy*.

The FAO approach, however, has its drawbacks. The more units at one level, the more difficult it is to remember relationships among them; for example, there are three orders that are characterized by a CEC of less than 24 cmol(+) kg^{-1} clay: Lixisols, Acrisols, and Ferralsols; and four orders are required to have a mollic epipedon: Greyzems, Chernozems, Kastanozems, and Phaeozems. The creation of a "superorder" category could highlight these relationships and help users understand them with less memorizing efforts. The developers of the new FAO legend have not made that step.

2. Soil Climatic Regimes

a. The FAO–UNESCO Approach

All units of earlier versions of the FAO–UNESCO legend that included climatic criteria have been redefined or dropped. There are, except when permafrost is present, no climatic criteria among the FAO–UNESCO differentiating characteristics anymore. All units are described in terms of permanent morphologic, chemical, or physical properties only. The trend to remove all attributes from soil classifications that cannot be identified at all times is also favored by French (Fauck *et al.* 1979; AFES, 1987) and Brazilian (EMBRAPA, 1981) soil scientists. Some national classifications, for example the Kenya soil survey (Sombroek *et al.*, 1982) have adopted the same approach.

The reasons given for excluding climatic attributes from soil classification systems relate to the difficulties in measuring soil temperature and moisture regimes and the fact that, consequently, many soils cannot be classified because of lack of information.

b. Advantages and Disadvantages

The avoidance of the difficulties in measuring climatic attributes of soils makes the FAO–UNESCO legend easier to use than *Soil Taxonomy*, which generally requires estimates of seasonal moisture conditions to allow the identification of a soil at the suborder level. There is a price to pay however: the FAO–UNESCO legend does not provide any information on soil climate, which, undoubtedly, is an important attribute with regard to soil survey interpretations, particularly on small-scale maps that the FAO legends are intended to serve.

It is true that in most temperate glaciated areas, climatic properties of soils can be inferred from morphologic characteristics. As a rule, young and incompletely weathered sediments react to environmental conditions in such a way that they show the marks of soil-forming processes imposed by climate. Mollic and umbric epipedons are examples. However, in the case of the tropics, which have not been submitted to rejuvenation by glaciers, many soil attributes have developed under climatic conditions that are different from the present ones. On the other hand, most constituents of strongly weathered soils do not react to current climate. Many pedological features in the tropics are relic. The FAO–UNESCO approach, which may be justified in temperate regions, fails when soils of old geomorphic surfaces in tropical areas are described in FAO–UNESCO terminology.

The developers of the new legend are aware of the fact that the elimination of climatic attributes causes serious losses of information. They consider this shortcoming can be corrected by combining soil information with agro-ecological zone maps, if available.

3. The Third-Level Category

The newly introduced third-level category of the FAO–UNESCO legend has not been fully developed. It only provides a mechanism to name groups of soils that intergrade either to first- and second-level taxa (e.g., Gleyi-Dystric Fluvisols), or can be distinguished on the basis of phase criteria (e.g., Fragi-Ferric Luvisols). Apparently, no mutually exclusive third-level taxa have been established.

The units proposed at the third level are intended to serve the specific needs of national or regional maps; it is accepted that further guidance will be needed to retain adequate uniformity of the system to permit technology transfer. Examples of third-level subunits exist for Kenya (Sombroek *et al.*, 1982) and the European communities (Commission of the European Communities, 1985).

B. Diagnostic Subsurface Horizons

All diagnostic subsurface horizons that are used at the highest level are mutually exclusive, although some of them are subdivisions of others. Two new "master" diagnostic horizons that are important in the classification of soils of the tropics have been introduced: the *ferralic* B horizon, to replace the oxic B horizon, and the *argic* B horizon, to replace the argillic B horizon. Note that the FAO–UNESCO legend does not clearly separate the concepts of genetic horizon designations (A, B, C nomenclature) from the concepts of diagnostic horizons.

The new legend has reversed the sequence by which the criteria related to illuviation (the argic or argillic B horizons) and weathering (the ferralic or oxic B horizons) are used in the key. Weathering now has precedence over clay illuviation. This was not the case in the 1974 legend (FAO, 1974), which followed more closely the *Soil Taxonomy* (Soil Survey Staff, 1975) decision sequence, where the argillic overruled the oxic horizon.

1. The Ferralic B Horizon

The revised FAO–UNESCO legend has modified its previous definition of oxic B to become *ferralic* B by introducing three new attributes. The intent is to create a horizon diagnostic of strong weathering.

First, FAO–UNESCO dropped the "apparent" qualifier in the CEC and ECEC requirements of the clay fraction. *Soil Taxonomy* calculates the apparent CEC of the clay by dividing the CEC of the fine earth fraction of the total soil by its clay content without correction for organic matter. This change is justified because the objective of the CEC criterion is to identify the type of clay mineral, and because contamination by organic matter obviously leads to erroneous interpretations.

Second, the *silt/clay* ratio, which should be less than 0.2, introduces a long-recognized index of weathering stages (Van Wambeke, 1959, 1962; Bennema and Camargo, 1979, Sombroek and Muchena, 1979).

The third important attribute of the ferralic B is its low content of water-dispersible clay. This characteristic is used by Brazilian soil scientists to identify the horizons in which clay translocation has practically stopped. It is related to the surface charge of the clay fraction and thus is indicative of weathering stages.

2. The Argic B Horizon

The changes in diagnostic horizon definitions in the FAO–UNESCO legend are more drastic than in *Soil Taxonomy*. The new keys to *Soil Taxonomy* did not change the definition of the argillic horizon, which has strong pedogenic implications that reflect clay illuviation. FAO removed the argillic from the system to replace it with an argic horizon that can be the result of many different soil-forming processes.

The new argic B horizon is not a *subsurface-higher-clay-content horizon* that is exclusively the result of illuviation. Other processes such as destruction of clay in the overlying horizon, selective erosion of clay, sedimentation, or biological activities may also lead to its formation. A mere lithologic discontinuity, however, is not sufficient. Clay skins are only required when no other layer than a plow layer overlies the argic B horizon.

The argic B horizon is exclusive of the ferralic; in other words, it should lack the set of properties that characterize the ferralic B horizon. The argic B needs to have a texture of sandy loam or finer.

There are no new horizons in the revised FAO–UNESCO legend that parallel the kandic horizon; neither are there any new criteria that are comparable to the *sharp* textural change criterion used at a very high level in the new 1987 *Keys to Soil Taxonomy*.

C. Changes to Taxa

Only the taxa that are considered important in the tropics are discussed in this section.

1. Plinthosols

The presence of more than 25% plinthite by volume within 50-cm depth or within 125-cm depth if underlying an albic E horizon is diagnostic for Plinthosols. This criterion is used in the key before any other major differentiating characteristic that is important for the classification of soils of the tropics; thus, it is considered by the FAO–UNESCO legend as being highly critical both for soil genesis and land use.

There is no equivalent taxon in *Soil Taxonomy* for the first-level FAO–UNESCO class. The *Keys to Soil Taxonomy* (Soil Survey Staff, 1987) generally consider plinthite at the great group level.

2. Ferralsols

After eliminating the shallow, organic, poorly drained, and sandy soils as well as the Vertisols, Podzols, Andosols, Solonchaks, and Plinthosols, the FAO–UNESCO key recognizes as Ferralsols *all* soils that have a ferralic B horizon.

The Ferralsols may have or may not have an argillic horizon or an argic B horizon within the pedon. The Ferralsols thus group all strongly weathered soils. Geographically, they probably cover areas of old geomorphic surfaces or soils derived from parent rocks that weather easily under high-temperature humid conditions. This classification agrees with the preferences of soil scientists who consider that the weathering criterion has more associated attributes that are important for the objectives of the classification than the textural differences produced by illuviation (Van Wambeke, 1979).

The second-level subdivisions of the Ferralsols are mainly based on the presence of plinthite (Plinthic Ferralsols), the delta pH value of the B horizon (Geric Ferralsols), the kind of epipedon (Humic Ferralsols), and the color of the B horizon (Rhodic or Xanthic Ferralsols). The identification follows a key sequence that ends with Haplic Ferralsols. Base saturation is only used in the case of Geric and Humic Ferralsols.

3. Soils with an Argic B Horizon

The soils without a ferralic B but with an argic B horizon could be considered under a kind of a superorder that in FAO–UNESCO terminology would include the *Nitisols, Alisols, Acrisols, Luvisols, and Lixisols*. The criteria that separate these classes are placed in the key according to this sequence:

First:		Thick, deep argic horizon with nitic properties	→	NITISOLS
Second:		Other soils, with CEC ≥ 24 cmol(+) kg^{-1} clay		
	and			
		base saturation (NH_4OAc) *less* than 50%	→	ALISOLS
Third:		Other soils, with CEC < 24 cmol(+) kg^{-1} clay		
	and			
		base saturation (NH_4OAc) *less* than 50%	→	ACRISOLS
Fourth:		Other soils, with CEC ≥ 24 cmol(+) kg^{-1} clay		
	and			
		base saturation (NH_4OAc) 50% or higher	→	LUVISOLS

Fifth: Other soils, with CEC < 24 cmol(+) kg^{-1} clay

and

base saturation (NH_4OAc) 50% or higher → LIXISOLS

Note the importance given to the nitic properties. Nitic properties include attributes such as structure ("polyhedric" or "nutty") and free iron oxide (Fe_2O_3) content, usually more than 12% on the clay. When they are associated with a thick argic B horizon, they overrule the other criteria related to the exchange properties of the clay. The transition between A and B horizons in Nitisols has to be gradual or diffuse. The Nitisols probably group most of the deep, not completely weathered soils developed on basic rocks in the tropics.

The Luvisols and Alisols characterize soils that are not completely weathered and contain appreciable amounts of 2:1 clay minerals; the Alisols probably occur most frequently in the most humid parts of the wet and dry tropics.

The Acrisols and Lixisols characterize materials that are more strongly weathered. Given the low CECs of the clays in these soils, it may be difficult to separate Acrisols from Lixisols with adequate precision by means of the base saturation.

4. Soils Without an Argic B Horizon

The key recognizes two second-level taxa that are important in tropical and equatorial regions but that have neither a ferralic nor an argic B horizon. The criterion used is a CEC (by NH_4OAc) of less than 24 cmol(+) kg^{-1} clay.

The two subdivisions are the Ferralic Cambisols, which may be comparable to the Tropepts in *Soil Taxonomy*, and the Ferralic Arenosols, which compare to Tropopsamments.

IV. Soil Classifications and Soils of the Tropics

Soil classifications have an important role to play in the dissemination of knowledge among all nations in the world. Most of them are the product of national soil survey schools and many have contributed to build systems with an international scope. The task of the developers of these global systems is not easy. Certain points that are called to the attention of soil scientists by the publication of the two updates are highlighted herein. They relate to the institutional structures that do not always facilitate the implementation of changes in the taxonomies, to the hierarchical levels reserved for differentiating characteristics, to the isolation in which soil science is placing itself by eliminating climatic criteria, and to the necessary separation between soil taxonomies and mapping legends. A final comment on how well soil taxonomists make soil science attractive to others closes this review.

A. Implementation of Changes

Classifications are made to serve specific objectives and one system cannot serve all purposes equally well; each classification has to respond to given requirements that may differ from one institution to another. Their mandates do not always allow changes to be implemented without major disturbances to their functions as soil survey or land use planning institutions.

1. Soil Taxonomy

Soil Taxonomy is constructed to serve soil surveys at *all* scales, from exploratory to detailed. The latter often include soil units that are precisely defined in terms of a large number of attributes. *Soil Taxonomy* is also bound to serve small-scale maps that only show broadly defined soil units. The developers of *Soil Taxonomy* did not want to construct two separate *taxonomic* systems for the two approaches to soil survey: exploratory and detailed. All criteria introduced at a high level continue to function down to the lowest category. The SCS placed the *hiatus* (discussed later) at the separation between *taxonomic* and *map* units.

Many of the properties used as criteria in *Soil Taxonomy* are important for land-use interpretations and evaluations. Some legal, economic, or social values may be attached to these properties or to the names of the taxa. Given the intent to have only *one* system, it is obvious that small changes in the definitions of an order or suborder may have large repercussions in lower level taxa that are mapped on detailed surveys. For these reasons the SCS is rather reluctant to introduce changes in the system that affect the classification of soils in the United States.

There are other reasons as well. One of them relates to the linkage between taxonomic and map units. The SCS uses the *same name* to designate the lowest level taxonomic unit (the soil series, e.g., "Lansing series") *and* to designate a consociation, which is a map unit (e.g., "Lansing soils"). Under this system a change in a taxonomic definition may result in a change in the name of the map units that contain the modified taxon. Updates of the taxonomic system automatically imply a revision of the surveys, most of which are published.

Looking back, it might have been preferable not to use the same names for *map units* and *taxonomic units* and not to link both reference systems so tightly. The separation between taxonomic and map units would have been more clear-cut and would have kept both reference systems completely independent. Consequently, it would have given *Soil Taxonomy* more flexibility to make changes.

If the increase in knowledge and the inaccuracies in mapping (soils are hidden below the surface) require some flexibility in soil definitions, the simultaneous function of a soil classification system to serve as a vehicle for technology transfer, on the other hand, requires strict definitions. The approach by the SCS to build a *crisp taxonomy* and to provide a flexible parallel reference system—the *kinds of map units*—to describe map units seems to satisfy both requirements (Van Wambeke and Forbes, 1986).

2. The FAO–UNESCO Classification

The FAO–UNESCO classification in 1974 was designed to serve only the legend of the 1:5,000,000-scale "Soil Map of the World." The 1988 revision, on the other hand, is designed to serve national 1:1,000,000-scale soil maps. There are no binding obligations with soil units that are mapped on detailed soil maps. The mapping units shown on the small-scale maps are essentially composite (associations, complexes, or undifferentiated groups) and their components are seldom delineated separately. The classification does not have to accommodate units or ideas about units that do not fit satisfactorily into the general scheme because there is no institutional framework to make changes mandatory at national levels. For these reasons, changes are easier to implement and the FAO–UNESCO system, at the three levels, has more freedom to accept them than *Soil Taxonomy*.

B. Ceilings of Independence

The concept of ceilings of independence was introduced by Cline (1949). It is closely related to the "kingdom" concept that recognizes groups of individuals that are so different that they cannot be subdivided by the same set of characteristics. An example in biology would be plants and animals. In soil science, organic and mineral soils illustrate the existence of kingdoms in the soil universe. Kingdoms and ceilings of independence are attributes of hierarchical systems as opposed to relational and network approaches to data management.

A differentiating characteristic is said to be used above, or outside, its *ceiling of independence* when it starts separating like things within a given group.

In *Soil Taxonomy* the use of base saturation to separate Alfisol and Ultisols is an example of a violation of the principle of ceilings of independence when this criterion is applied to distinguish soils formed in materials with extremely low CECs. The criterion is said to work satisfactorily in the United States to separate soils according to the former concepts of Gray-Brown Podzolics and Red-Yellow Podzolics. It does not when it is used in strongly weathered soils of the tropics, which often have extremely low CECs. Van Reeuwijk, for example, (1984) compared the separation of certain soils on the basis of saturation percentages to playing dice.

The ceiling-of-independence problem can be avoided either by lowering the categorical level of the overused criterion or by raising the level of other criteria to remove certain soils from the group. In the example given earlier, raising the weathering stage and/or the CEC above the category where base saturation is used could solve part of the problem.

Ceilings of independence are also important in determining the relative categorical levels of criteria related to weathering and clay illuvation. The new *Keys to Soil Taxonomy* (1987) have upgraded the soil properties that are the result of strong weathering (the oxic horizon) above the argillic horizon, except for soils that have a *sharp* clay increase below a topsoil with less than 40% clay. They have, thus, partly corrected the imbalance between these two criteria.

The revised FAO–UNESCO legend has been more drastic in its changes by removing both the argillic and the oxic B from their classification. They have now in all cases given precedence to soil attributes indicative of strong weathering (the ferralic B horizon) above those that are marks of clay illuviation (the argic B horizon).

C. Use of Soil Climate in Soil Classification

The present trend in most classification systems, except *Soil Taxonomy*, is to delete all climatic criteria from soil classifications. Only chemical, physical, and other immediately recognizable properties are used to identify soils. This "pedocentric" approach is illustrated in the new version of the FAO legend (1988).

It may be true that in certain circumstances time constraints may force soil survey organizations to postpone or temporarily ignore soil climate. Soil climate is, however, an important agronomic and pedogenic factor. In geographic representations, it is extremely powerful especially on small-scale maps.

There are several reasons to regret the deletion of climatic soil attributes from soil classifications. First, it drastically reduces the usefulness of soil surveys for agronomic interpretations. Climate is essential in the selection of crops and the choice of management practices. The argument that agro-ecological zone maps fill the gap does not hold when these maps are not available or are published at inadequate scales, as in the case of the 1:1,000,000 map of the EEC (Commission of European Communities, 1985). There is no way one can recognize on the European soil map broad soil areas that are suitable for the major crops grown.

The simple overlaying of agro-ecological maps and soil maps is misleading in the case of the semi-arid tropics. Runoff and inflow of surface water that depend on the concavity or convexity of the slopes change the moisture regimes of many soils in these areas to extremes that cannot be inferred from the agro-ecological zone map. Smith (1986) made similar comments during his interviews on the rationale for concepts in *Soil Taxonomy.*

The second reason for regretting the omission of climatic attributes is that it discourages soil survey investigators to engage in more research on soil climatic regimes. What would plant scientists do if they only took into account plant characteristics that are always present and ignore flowers and leaves? Soils as "things out of doors" have their seasonal cycles that are extremely important for land use and soil genesis. There is a need to intensify the research that would allow better evaluation of soil climate and make soil survey more useful.

The weak efforts in soil climatic research by soil scientists is not limited to FAO–UNESCO. *Soil Taxonomy* has not changed any climatic criteria since 1975 and has contributed very little to increasing the interpretive effectiveness of soil moisture and temperature regimes. The moisture regimes are only intended to describe in very general terms the soil climatic factors that influence the dominant pedogenic processes that take place in the moisture control section. For this limited objective and where soils have been surveyed and observed during several years, this approach gives satisfactory results. However, if such information from surveys is not available, as in many tropical areas, more sophisticated simulation models than the Newhall (1972) model will be necessary to allow better estimates of climatic conditions. The Newhall model has some of the same shortcomings as the FAO–UNESCO approach: it does not take into account runoff or inflow and completely ignores soil physical properties.

The same remarks may be made on soil temperature regimes. Their definitions are, in most cases, only based on *mean annual* temperatures; although seasonal temperature variations are not so critical in the tropics as in temperate regions, investigations in maximum daily temperatures could add considerable interpretive value to the classifications.

D. Testing Changes to Classification Systems

There are no well-established methods to measure the value of changes that are introduced in soil classification systems. Some possible rationales are those that relate them to purpose; or increase the information content of the taxa, particularly for interpretations; or improve the understanding of the relationships among the units that are classified. A less subject-matter-oriented criterion is the consistency of the logical framework upon which the classification is built.

In most cases, judgments rely heavily on the opinions of the individuals who developed or used the system and who base their evaluation on the degree the updates satisfy their perception of the soil universe. Inevitably subjectivity resulting from views limited to certain regions is inherent to the testing methodology.

There is no published information of how the new introductions have been evaluated. The new FAO legend, to our knowledge, has not been used yet in actual mapping operations; neither have the new *Keys to Soil Taxonomy* been used in published soil survey reports. There probably isn't any method to estimate the value of theoretical taxa for which no representative profiles in the field have been observed.

The soil science community would obtain additional benefits from the updates if they would include listings of the new classifications of benchmark profiles, that is, those that have been published in earlier versions or that have been described and analyzed during the national or international soil correlation workshops. It would create more opportunities for soil scientists to participate in the validation process of the updates.

E. The Soil Survey Hiatus

There is almost never perfect coincidence between the boundaries of soil bodies that follow the conceptual limits of taxa and the field boundaries of soils as they are shown on soil maps. These discrepancies lead to several classification concepts, such as the "hiatus" by Butler (1980). There will always be, at least, one hiatus between conceptual classes in taxonomic systems and real soil bodies in the field. There can be additional hiatuses that can then be placed at distinct locations in the soil survey procedures, for example, at the separation between taxonomic systems for large-scale maps and small-scale maps. This would be the case when a given soil series of a local soil survey would be loosely linked to a great group of a more general national soil classification system.

A classification system is said to have no hiatus when the criteria continue to be valid from the highest level down to the lowest level and when all the classes at the same categorical level are mutually exclusive. These systems result in unambiguous and reproducible definitions of taxa that fit into a hierarchical framework. A hiatus is created when, at a certain level, a criterion that is used higher in the hierarchy does not separate objects at lower levels. Several structures are possible. For example, a soil series could belong to several great groups if it shared properties with several of them and, although each category may only contain mutually exclusive classes, no mutually exclusive linkage with a higher taxon would be requested. Other structures that could be adopted by soil classifications could be grouped into three data-management approaches: the relational, the hierarchical, or the network approach (Date, 1981).

A group of French soil scientists, although they have not discussed applications of their proposal to tropical soils (AFES, 1987), suggest a fuzzy set approach – (*grands ensembles à limites floues*, (*GELF*). Baize and Girard (1987) consider it unreasonable to require a strict classification of a profile and prefer an "association" approach (*rattachement*) to one or several previously established GELFs. They object to a soil classification that requires mutually exclusive classes. The system that defines the various GELFs is not described in the publication (AFES, 1987).

The SCS placed the hiatus between the taxonomic and the map unit definitions. It maintains a crisp taxonomic system with consistent criteria and mutually exclusive classes at all levels, including the lowest one, the soil series. It allows flexibility only in the definitions of map units, the definitions of phases, area distinctions, and miscellaneous land types that are used as additional reference systems (Van Wambeke and Forbes, 1986).

The new version of the FAO–UNESCO legend has not fully developed its third category but states that the "definitions of subunits should not overlap and should in no way conflict with the definitions of the soil units at first and second level" (FAO, 1988). There are no guidelines, however, on how large-scale soil units such as soil series would be linked to the three-level FAO–UNESCO classification.

Simplifications that include the elimination of lower level taxa always cause losses of information. In most soil classifications, these lower level units contain information for practical interpretations, such as texture, mineralogy, and so on, that may be as important as the higher level criteria. This is why a comprehensive system that considers all levels is the most effective for technology transfer.

However, depending on the map scale and the degree of detail required by interpretations, international correlation of soils could look into some of the data-management approaches to build less rigid systems than those imposed by hierarchical structures. How useful a fuzzy set approach without mutually exclusive classes could be for large-scale maps needs to be investigated.

F. Future Developments

1. Use of International Systems

Soil Taxonomy will probably remain the most frequently used international reference system in the Western world because it provides taxa that go down to the level of detailed surveys. It also provides a wealth of criteria and serves as a checklist of soil attributes to be observed when making soil surveys.

National surveys, however, may be inclined to use *Soil Taxonomy* only down to a certain level and then construct their lower categories according to their own needs. Certain subdivisions that may be useful in the United States may be irrelevant in other countries. The existence of a national soil classification does not exclude an international one to serve as reference system when there is a need for technology transfer from one country to another.

National systems may prefer the FAO–UNESCO legend as a reference base and build their lower category subdivisions according to relevant criteria in their nations. If soil moisture and soil temperature conditions in the national survey areas are relatively uniform, the lack of soil climatic information will not be felt. If extrapolation to different ecological zones is needed, the soil maps used without careful consideration of soil climate will have little interpretive value.

The revised FAO legend, which is based on a broader based "tropical" experience than *Soil Taxonomy* and is not tightly linked to an institutionalized framework, may be more attractive to developing nations located in the tropics. The value of the new criteria, however, will only become apparent after they have been used during actual soil survey operations in several parts of the world.

2. Contributions of Soil Classifications to Tropical Agriculture

The progress any science makes, particularly when contributions to it come from several parts of the world and when on-site observations are a *conditio sine qua non*, depends to a large extent on the quality of its technical language that constitutes its basis of communications.

Although it is not the responsibility of soil classifiers alone to give in their publications adequate references to the kinds of soils that are used in agricultural activities, soil classification can play an important role on the international scene to facilitate exchange of scientific information. At present, the abstract journal *Soils and Fertilizers* organizes its "soil types" subdivisions according to various criteria among which some are anthropogenic, chemical, climatic, ecologic, genetic, lithologic, mineralogic, palaeosolic, physiographic, or textural. The genetic soil types are organized according to the first level of the FAO–UNESCO 1974 classification.

A comparison with what other disciplines have achieved in the development of a technical language may serve as a yardstick to measure where soil science stands. Considering the chemical substance index names that the Chemical Abstract Service of the American Chemical Society uses to organize their bibliographic data base, one reaches the conclusion that soil taxonomists still have a long way to go to contribute in the same proportion to the development of soil science. It should also stimulate soil scientists to increase their efforts in refining the systems or the taxonomists to improve their methods for reaching that goal.

A recommendation: since the updates often change definitions without changing names, users of the systems are invited to give the date of the revision they refer to. A 1983 Oxisol is not necessarily a 1987 Oxisol, and a 1983 Ultisol may now belong to the 1987 Oxisols. This will reduce the risks of confusion in the transfer of technology.

A final comment: progress does not depend only on systems. People who practice the soil science discipline or borrow from it are more important. The task of the teachers of soil science who train students interested in international agricultural development is not made easier by the drifting apart of the FAO and *Soil Taxonomy* approaches.

References

AFES, Association française pour l'étude du sol. 1987. "Référentiel pédologique Français, 1ere proposition." INRA. Plaisir, France.

Baize, D., and M.C. Girard, 1987. Mode d'emploi du référentiel. In: AFES, Association française pour l'étude du sol. 1987. "Référentiel pédologique Française, 1ere proposition." 5 pp. INRA. Plaisir, France.

Bennema, J., and M. Camargo. 1979. Some remarks on Brazilean Latosols in relation to the Oxisols of *Soil Taxonomy. Proceedings of 2d International Soil Classification Workshop.* Part 1:233–261. Bangkok, Thailand.

Butler, B.E. 1980. *Soil classification for soil survey.* 129 pp. Oxford University Press, NY.

Cline, M.G. 1949. Basic principles of soil classification. *Soil Science* 67:81–91.

Commission of the European Communities. 1985. "Soil map of the European communities. Scale 1:1,000,000." Office for Official Publications of the European Communities, Luxembourg.

Date, C.J. 1981. *An introduction to database systems.* 3 ed. 574 pp. Addison-Wesley, Reading, MA.

EMBRAPA, Empresa Brasileira de Pesquisa Agropecuária. 1981. "Sistema Brasileiro de classificação de solos." (2ª Aproximação). 107 pp. Rio de Janeiro, Brasil.

FAO. 1974. *FAO–UNESCO soil map of the world. 1:5,000,000.* Vol. 1, *Legend.* UNESCO, Paris.

FAO. 1988. *FAO–UNESCO soil map of the world. Revised legend.* World Soil Resources Report 60. FAO, Rome.

Fauck, R., M. Lamouroux, A. Perraud, P. Quantin, P. Roederer, J. Viellefon, and P. Segalen. 1979. *Projet de classification des sols.* 301 pp. ORSTOM. Bondy, France.

Moormann, F.R. 1975. Circular letter no. 1. In: F.R. Moormann, ed. "Excerpts from the circular letters of ICOMLAC." pp. 2–10. SMSS Technical Monograph No. 8. 228 pp. Soil Management Support Services. Soil Conservation Service. USDA. Washington, DC.

Newhall, F. 1972. *Calculation of soil moisture regimes from the climatic record.* Soil Conservation Service. USDA. Revision 4. Washington, DC.

Smith, G.D. 1986. "The Guy Smith interviews: rationale for concepts in Soil Taxonomy." SMSS Technical Monograph No. 11. 259 pp. Cornell University, Ithaca, NY.

Soil Survey Staff. 1983. "Keys to Soil Taxonomy." SMSS Monograph No. 6. 244b. Cornell University, Ithaca, NY.

Soil Survey Staff. 1987. "Keys to Soil Taxonomy." SMSS Monograph No. 6. 3d Printing. 280 pp. Cornell University, Ithaca, NY.

Soil Survey Staff. 1975. *Soil taxonomy: A basic system of soil classification for making and interpreting soil surveys*. Agriculture Handbook No. 436. 754 pp. Soil Conservation Service. GPO, Washington, DC.

Sombroek, W.G., H.M.H. Braun, and B.J.A. Van der Pouw, 1982. "Exploratory soil map and agro-climatic zone map of Kenya, scale 1:1,000,000. Exploratory Soil Survey." Report No. El, Kenya Soil Survey, Nairobi.

Sombroek, W.G., and F.N. Muchena. 1979. LAC Alfisols and Ultisols in eastern Africa; some problems to their identification and classification. *Proceedings of 2d International Soil Classification Workshop*. Part 2:129–138. Bangkok, Thailand.

Van Reeuwijk, L.P. 1984. "Laboratory methods and data exchange program for soil characterization. A report on the pilot round. Part II: Exchangeable bases, base saturation and pH." Technical Paper No. 8. 28 pp. ISRIC. Wageningen. The Netherlands.

Van Wambeke, A. 1979. Concepts of Oxisols. Report on results of questionnaire. *Proceedings of 2d International Soil Classification Workshop*, Part 1: Malaysia. 25–36. Bangkok. Thailand.

Van Wambeke, A. 1962. Criteria for classifying tropical soils by age. *J. Soil Science*. 13:124–132.

Van Wambeke, A. 1959. Le rapport limon/argile, mesure approximative du stade d'altération des matériaux originels des sols tropicaux. CCTA. *Comptes rendus de la III Conférence interafricaine des sols*. Vol. 1. pp. 161–168. Dalaba, .

Van Wambeke, A., and T. Forbes, eds. 1986. "Guidelines for using *Soil Taxonomy* in the names of map units." SMSS Technical Monograph No. 10. 75 pp. Cornell University, Ithaca, NY, and Soil Management Support Services, Soil Conservation Service, USDA, Washington, DC.

Index

MIX
Papier aus verantwortungsvollen Quellen
Paper from responsible sources
FSC® C105338

If you have any concerns about our products,
you can contact us on
ProductSafety@springernature.com

In case Publisher is established outside the EU,
the EU authorized representative is:
Springer Nature Customer Service Center GmbH
Europaplatz 3, 69115 Heidelberg, Germany

Printed by Libri Plureos GmbH
in Hamburg, Germany